Ten Commandments
of Better Contracting

Also of Interest

How to Work Effectively with Consulting Engineers: Getting the Best Project at the Right Price (ASCE Manuals and Reports on Engineering Practice No. 45, updated edition). 2002. Stock # 40637

Quality in the Constructed Project: A Guide for Owners, Designers, and Constructors (ASCE Manuals and Reports on Engineering Practice No. 73, second edition). 2000. Stock # 40506

Surety Bonds for Construction Contracts, edited by Jeffrey S. Russell. 2000. Stock # 40426

Construction Contract Claims, Changes and Dispute Resolution, second edition, edited by Paul Levin. 1998. Stock # 40276

Capturing Client Requirements in Construction Projects, edited by John M. Kamara, Nosa F. Evbuomwan, and Chimay J. Anumba. 2002. Stock # 3103

Ten Commandments of Better Contracting

A Practical Guide to Adding Value to an Enterprise through More Effective SMART Contracting

Francis Hartman

Published by American Society of Civil Engineers

Library of Congress Cataloging-in-Publication Data

Hartman, Francis T., 1950-
    Ten commandments of better contracting : a practical guide to adding value to an
    enterprise through more effective smart contracting / Francis T. Hartman.
        p. cm.
    Includes bibliographical references and index.
    ISBN 0-7844-0653-7
    1. Contractors' operations--Management 2. Civil engineering contracts. I. Title

    TA210.H28 2003
    658.4--dc21

                              2003052470

Published by American Society of Civil Engineers
1801 Alexander Bell Drive
Reston, Virginia 20191
www.asce pubs.asce.org

Contents

Chapter 8   Thou Shalt Not Mess Unduly with the Contract After it is Agreed

Chapter 9   Thou Shalt Deal Rationally and Fairly with Inevitable Changes to a Contract

FIGURES

TABLES

Acknowledgments

This book is the result of a lot of work and inspiration by many people. It contains gems from research at the University of Calgary funded through my Chair and supported by the Natural Sciences and Engineering Research Council, the Social Sciences and Humanities Research Council, more than 40 industry sponsors and the Canadian Project Forum.

Research contributions and ideas from my students included significant work by Patrick Snelgrove, Cheryl Semple, Zainul Khan, Keith Pedwell, Roch DeMaere, Mona Sennara, Liwen Ren and Ramy Zaghloul. Many other students also contributed ideas and were generous in allowing me to complete the book by giving up time we would otherwise have spent together.

Colleagues and friends in industry contributed ideas, stories and advice that have been woven into this work. Some of the more significant contributors include my good colleague and academic partner George Jergeas and my business partner Tom Gosse. Others who lent ideas and advice include Bob McTague with his seemingly endless wisdom, Mike O'Neil with his insight and understanding of the human condition and Chris Levy for reviewing the legal points in Chapter 1 for accuracy.

On the production side, thanks to Janine Revay for all the hard work in proofing and formatting the book and to Rafi Ashrafi for the hard work in assembling the references and bibliography that support and add to the content. The reviewers, James E. Laughlin and Paul Levin, contributed advice, suggestions and constructive criticism that led to a stronger document. Any remaining mistakes are entirely my responsibility.

Last but not least, thanks to my family: Margaret and the children, who patiently grew older as I became harder to live with through the production of this work. Thanks Tamsin, Richard, Christopher and Kirsten.

Introduction

Contracts, to me, are a bit like a marriage. In fact, some people are going the other way and making their weddings more like a contract. Not all of our contracts (or marriages for that matter) are successful. In North America, contract lawsuits are the second largest contributor to private litigation, only topped by personal injury cases. We add little to our community, our corporate bottom line, the value of our products and services or our competitive edge by suing each other. I hope that this book helps us increase value to our own organization and to that of our business partners by reducing conflict and increasing performance. We need to work harder at making our contracts more successful.

This book is about getting more out of our contracts – whether we are buying or selling goods or services. Its 12 chapters were written with the busy practitioner in mind. I hope you will find it easy to read or skim through. Although much of the content is based on contracting in the construction industry, the underlying principles apply to other businesses too. For this reason, you will notice, particularly in the earlier chapters, that construction-specific examples are scarce or missing.

The work is based on a mix of research, hard-won experience and some trial and error in testing contracting innovations. Frankly, I do not think I go far enough with wacky ideas in this book, because I believe there is enormous scope for improving how we manage our buying and selling in business. However, all of the cases and examples (even the hard-to-credit ones!), are based on real situations. Only the identities of the players have been changed. The book offers a compromise between what probably could be done and what is prudent to suggest and actually have accepted by practical business people. That said, everything in this book has been tried successfully on real projects.

An experienced procurement specialist, project manager or other related professional will recognize some – or even many – of the ideas presented. If so, good on ya!! You are doing a great job. If not, please consider the ideas. They have served others well.

The content is aimed at setting ten commandments for more effective contracting and then putting those commandments into a cohesive package that we can apply in the workplace. The approach is a framework. It is not prescriptive. You know

Each of the commandments is explained in the first ten chapters. Each chapter sets out the problem and offers some immediate suggestions that might help improve performance on your next contract.

Finally, I know that I tend to use an intellectual cattle prod. I do make some pretty bold statements (which may be surprising or get some readers riled up or upset). But then I explain what I have said, and I hope readers will come along for the ride and get something out of the experience. This book is designed to get us thinking about the real issues that plague our projects and are so insidious that we either do not give them thought time or we accept them as "normal."

Enjoy the book. Best wishes to you, and may the book contribute to many future business and contract successes.

Terms You Should Know

Following are a few words that have a particular meaning or that are used in a specific sense in this book. Words in *italics* are ones that are defined elsewhere in this section.

Agent: An individual or *enterprise* acting as a fully authorized representative of another enterprise.

Agreement: See *Contract*.

Bid: (noun) An *offer* by a *vendor* to provide *goods* or *services* to a *buyer*.
(verb) To make an *offer* for *goods* or *services*.
See also *Tender*.

Blue Trust: The type of *trust* that is based on the competence of the trustee as perceived by the trustor. See also *trust*.

Buyer: An individual or *enterprise* or someone acting as *agent* for an enterprise that is acquiring or intending to acquire *goods* or *services*. Also referred to as *owner*.

Charter: A document that fully describe the objectives, constraints and other relevant criteria for a *project*. The charter represents the license of the team spending the *buyer's* money to expend those funds for a project.

Common Law: The legal framework for precedent-based law that originated in Great Britain and that forms the basis of the legal system in many countries, including Australia, Canada, New Zealand, South Africa, the United Kingdom and the United States.

Condition: A term of a *contract* that places specific and binding requirements on one or more of the *parties* for *performance* under the contract.

Contract:	A legal binding and enforceable *agreement* between two or more parties under which there is an obligation to exchange value – usually money, in exchange for *goods* and *services*.
Consultant:	*Designer* or other professional adviser.
Contractor:	A *vendor* of *goods* or *services*. Also referred to as a *vendor*.
Cost Plus:	A type of *contract* payment method by which the *contractor* is compensated for the actual cost of the *work* plus a predetermined fixed, variable or percentage *fee*.
Designer:	Architect, Engineer or other professionals involved in design of a project to be built by the *Contractor*. *Designers* may also be *Contractors* or *Subcontractors*.
Enterprise:	Any individual or group that provides *goods* or *services* to others. Typically a legal entity, an enterprise may be privately owned, may be publicly traded or may be a government agency or a not-for-profit organization.
EPC:	*E*ngineering, *p*rocurement and *c*onstruction; usually refers to an integrated service providing all three components of these *goods* and *services* through a single *enterprise*.
EPCM:	*E*ngineering, *p*rocurement and *c*onstruction management. See also *EPC*.
Fee:	A charge by a *vendor* to cover overhead, profit and other *services*.
Fixed Price:	A type of *contract* payment method that sets the *contract price* at the outset, often as a result of the *buyer* accepting a *bid* from a *contractor*.
Goods:	Product, materials or commodities, including specialist equipment. Any physical product may be included in this definition.

Lump Sum:	A type of *contract* payment method by which the *vendor* is paid on completion of its *contract*. Often incorrectly applied to *fixed price* or *stipulated price* contracts.
Offer:	(noun) A legally binding proposal to another *party*. (verb) The process of presenting a proposal for acceptance by another *party*.
Owner:	See *Buyer*.
Party:	Any *enterprise* that is bound by a *contract*.
Price:	The offered and accepted value of a *contract* for supply of *goods* or *services*.
Priority Triangle:	A graphic of an inverted triangle that is used to help align stakeholders' priorities in terms of cost, time and performance on a *project*.
Profit:	The net income that exceeds all costs associated with generating the total income.
Project:	An endeavor with a clearly defined end result, objectives, and limited resources.
Project Charter:	See *Charter*.
Red Trust:	The type of *trust* that is based on the intuition and possibly the emotional response to the trustee as perceived by the trustor. See also *trust*.
SBS or SMART Breakdown Structure:	A modified work breakdown structure that identifies the *project* mission as the top element and the key results expected by identified stakeholders at the second level.
Services:	Expertise- or knowledge-based service, such as design, supervision or other service such as rental of labor or equipment or use of infrastructure. Differentiated from *goods* in that the acquired product is not tangible or is temporary.
Subcontractor:	A Contractor performing work for the *Contractor*.

Three Key Questions:	Three questions that are used to help identify the critical stakeholders, how they will measure success of a *project* or *contract* and when that measurement will occur (usually the defined end of the *project* or *contract*).
Tender:	An *offer* by a *vendor* to supply *goods* and *services* to a *buyer*. See also *Bid*; terms may be used interchangeably.
Third Party:	Any individual or *enterprise* that is not a *party* to a *contract*.
Trust:	Confidence and belief in the capability of others to meet our expectations. See also *blue trust, yellow trust* and *red trust*.
Unit Rate:	A type of *contract* payment method by which the *contractor* is compensated for the quantities actually delivered based on a per unit rate established in the *contract*.
Vendor:	See *Contractor*.
Warranty:	A guarantee of performance to specifications of the *goods* or *services* provided under the terms of a *contract*.
Work:	The *goods* and *services* plus any ancillary effort required to deliver what is required under the terms of a *contract*.
Yellow Trust:	The type of *trust* that is based on the consistent integrity of the trustee as perceived by the trustor. See also *trust*.

Icons

Throughout this book, you will find mini summaries of salient points. These are in bold and are marked with one or more icons that are described below.

This could make your life easier.

This is a hazardous thing – pay attention!

This represents a bright idea – consider it in the future.

This could save you time.

This is something you may want to read more about – check out the bibliography.

THOU SHALT CONTRACT WITHIN THE LAW AND THE WORKING ENVIRONMENT OF THE CONTRACTING PARTIES

> *Typical Contract Problems and their Causes: Common Law and Tort Issues, Changes in the Contracting Environment*

Chapter 1 has three main parts. The first part is a high-level summary of contract legal issues in common law. Contract formation, breach, frustration, misrepresentation and other basics are covered, as are obligations under tort. Layman's terms are used throughout, with the focus on setting the historical and legal context for the rest of the book. The second part of Chapter 1 addresses the business, technical and social context for the discussion that follows. The third part looks at how changes have led to real opportunities for better contracting. We set the stage for considering better ways of doing business with our customers and suppliers through more effective contracts and contracting practices.

Common Law is the law established by precedent. It stems from the judicial system of the British Empire. As a result, it remains the foundation of the legal systems of Canada, the United States, Australia, New Zealand, South Africa and, of course, the United Kingdom amongst others.

Chapter 1 covers the essentials of contract and tort laws in common law jurisdictions – without your having to become a lawyer first.

I didn't realize that the Client was
THIS serious about Partnering!

Whereas the first part of Chapter 1 is material that most readers will be familiar with, the second part is a bit different. This second part focuses on the working environment and how it has changed in the past few decades, pushing it ahead of where contracting practices are today. That contracting practices, cases and habits are asynchronous with the rest of the working environment we are in is symptomatic of the conservatism with which we address the business of buying goods and services.

The second part of Chapter 1 explains why our conservative approach to procuring goods and services has left many of us behind, missing opportunities while we "play it safe".

1.1 Contracting Basics

In Section 1.1 of Chapter 1, we look at the basic legal framework for contracts in common law jurisdictions. This is not a legal treatise, so don't treat it as such. Each local jurisdiction has its own twists and turns, so it pays to get specific legal advice from your tame lawyer for each situation.

Chapter 1 provides only an overview of the law. As the law varies from one jurisdiction to another, consult a lawyer for advice on your particular situation, if you have a need.

1.1.1 Contract Law as Part of Common Law

Common law is based on precedent. It stems from a long history of decisions made in the court of the king or other friendly dictator. When the disputes became too frequent for the king to be personally involved in every case, he assigned others to be judges, and they sat on the king's bench to represent him. So as not to make bad judgments, whenever possible, they went back to what had been decided by the king on previous occasions. And that is where, in reader's digest form, the concept of legal precedent comes from. So, in common law, we have two main branches outside the world of criminal law and legislation. These are the two areas in which we resolve disputes between people. One is the law of tort and the other is contract law. In tort, anyone can sue anyone else if they feel that the other party has damaged their fundamental right as a citizen to enjoy their life and property. The concept behind this is simple: We each owe others a duty of care, so that we do not harm each other. If we do harm, then we need to make things good again. In other words, there is a "social contract" that we all live by in society.

Contract and tort laws are based on precedent.

No, he's not a lawyer, that's his wife, Susan.

In addition to this, we enter into legally binding agreements to do things for each other. These agreements are contracts. They must comply with some sensible principles in order to be legal and binding. These sensible principles have evolved, and, in some cases, become a bit less than sensible in places. Let's take a look at them.

1.1.2 Contract Formation

A contract cannot exist unless there is intent to have a deal of some sort. So the first thing we look for is intent to be legally bound to perform some duty or provide some materials.

 The first ingredient in a contract is intent to be legally bound.

Being normal and rational beings, we usually expect a contract to involve an exchange of value. This is the second ingredient we look for. Interestingly, there is no obligation to have an equitable exchange of value. So, if you are a buyer, you can reasonably expect some value for your money, but it is up to you to negotiate, haggle or deal for that value. This "value" is often referred to by lawyers, law books and in legal circles as "consideration".

The second ingredient in formation of a contract is an exchange of value (which needs to be convertible to a financial equivalent) between the contracting parties.

This is a reasonable reason for limiting the people who can enter into a contract to those who are older than the age at which they are legally considered to be a minor and those who are in a position to represent the organization (if we are dealing with a company or agency in this contract). Come to that, whatever we undertake to do or provide needs to be legal for the contract to be enforceable in law. For example, contracting with someone to rob a house is not an enforceable deal in the courts of law.

A contract needs to be between people or organizations that are legally competent to enter into the deal.

To be enforced in a court of law, a contract needs to be for something legal.

Finally, we need the basis of the deal, so we need to have someone offer you something and we need to have someone else accept the deal. The terms and conditions are those declared, negotiated and agreed in the process of entering into the contract. Making a counter-offer does not mean the same as acceptance. We need to be sure that after the haggling is over, one party must make an offer and the other must accepts it without changes.

A contract needs someone to make an offer for goods and services and someone else to accept the deal.

If I were to offer you a car for sale at $5,000.00, that would be an offer. If you said you would buy it for $4,800.00, that would **not** be acceptance. It would be a counter-offer. If I accepted that price, but added the condition that the car would be sold "as is" with no comeback to me at that price, that too would be a counter-offer. Acceptance means accepting the terms and conditions exactly as set out by the other party.

 Acceptance of an offer must be acceptance of the offer exactly as made. If the terms change, it becomes a counter-offer.

The terms and conditions of the contract are easier to manage if they are explicit and clear. And they do not even have to be spoken, let alone be in writing. Let's look at an example in Cut Me a Deal.

Cut Me a Deal

There is a sign in a barber's shop that says "Haircuts: $20". You walk in and sit down in one of the chairs. The barber comes along and, without a word, wraps you in one of his hair-catcher sheets. He starts cutting your hair. Still no word from either of you. Do you have a contract? The answer is yes.

You saw the sign. This was the offer. You sat down and did not do anything when the barber put the sheet over you and started cutting your hair. That was acceptance of the offer. And you did not move as the barber continued cutting your hair. These actions – or inaction – represent an intent to be bound. You get a haircut. The barber gets $20. There is an exchange of value. Provided both you and the barber are not minors or of unsound mind, all of the ingredients for a contract are present. The terms and conditions are simple. Exchange a haircut for $20.

Even the smallest transactions of daily life are often tied into contracts in their simplest form. Normally these contracts do not cause any problems because they just happen. However, sometimes things do go wrong. If the "wrong" is related to failure of one party to a contract to do what they promised, then this is a breach of contract.

We routinely enter into contracts without a second thought. They rarely go wrong. When they do, we see the results in the form of apology or compensation. Occasionally, they end up in litigation.

1.1.3 Breach of Contract

There are many ways that contracts can be breached. Generically, breach of contract is the result of a failure of one party to the contract to perform or meet its responsibility. This can take many forms, including the following:

- Failure to make a payment
- Not delivering what was contracted for in quality or quantity
- Making a mistake that damages the other party through error or omission
- Late delivery if a date or urgency is a term of the contract.

A contract is breached when one party fails to provide what is required of them under the terms of the deal.

This type of breach allows the damaged party to recover its position. Such a position can be defined as if the parties had either not entered into the contract or they have been put back in the position that they would have been in had they received what they should have if the contract had not been breached. (I hope you do not have to read that sentence too many times to get the intent!)

Breach of a contract will likely damage the party that relied on the contract to achieve something.

Generally, contract law provides for financial consideration to the party damaged by such a breach. While the value of the compensation is an estimate (and sometimes a very rough estimate!) of the cost to make good the breach, it may approximate the damaged party's view of what it would really cost for full recovery.

The injured party in the deal is normally entitled to recover the damage done.

This principle of recovery from a situation – in order to be made whole – applies generally, but the real value of the damages that result from a breach is often difficult to prove. Because of this, the concept of liquidated damages has evolved. Liquidated damages quantify the expected damages up front and write this quantity into the contract. Thus, the damages are agreed upon at the outset. Typically, these damages are identified for the buyer, but not for the seller. The main reason for this is that it is the buyer who normally writes the contract!

Liquidated damages are a way of pre-agreeing to the values of specific breaches of contract.

1.1.4 Terms and Conditions

Liquidated damages are just one common term in a contract. The terms and conditions of a contract set the rules for business between the two contracting parties. Very often they contain clauses that are designed to stop things from going wrong for the buyer (or for the seller if it gets to participate in formulation of this part of the contract). Over time, and as things have gone wrong, buyers have developed more clauses to cover these eventualities. We will come back to the implications of this process.

Terms and conditions of a contract set the enforceable business rules for how the parties to the contract behave.

1.1.5 Typical Content of a Contract

In addition to the terms and conditions of a contract, there are typically a number of other components. We will look at these now, and define what we mean by each of the terms describing these parts.

Agreement: This is the primary document. It is the one that is normally signed by the parties to the contract. It will typically include, by reference, all of the other documents that make up the contract. Normally, if there is a discrepancy between this primary document and any other documents included with or referenced by the agreement, the primary document will govern. It is not unusual for the contract to lay out the priorities of one document over another. Most commonly this is done in the General Conditions.

General Conditions: The General Conditions provide the basic commercial and legal terms for the contract. They will describe payment mechanisms, insurance coverage, transportation of goods, who owns intellectual property produced under the terms of the deal and a myriad of other details of this type. The General Conditions are normally the second most important document, after the Agreement. If there are any differences between this document and others, this document will be interpreted as the correct one. Only the Agreement will normally govern over the General Conditions. Most industries offer one or more standard forms for General Conditions. Large organizations also have their own standard forms. This said, there are often situations that require modifications or additions to such standards. To address these situations, Special or Supplementary Conditions of contract have evolved.

In law, there is an important distinction between a contract term that is a Condition and one that is a Warranty. This becomes important if the term is breached because recourse of the injured party is different. If a Warranty was breached, then the damaged party must still continue with its part of the contract, but it can recover damages later. If, however, it is a Condition that is breached, then the damaged party has a choice. It can terminate the contract or continue – basically retaining the right to choose whichever is most advantageous. Either way, it remains entitled to compensation. So it pays to write a contract term as a Condition rather than a Warranty if you want this choice in the event of a breach. Particular conditions and rules may apply that change this general guideline in specific situations or with specific clients (such as government agencies). Professional legal advice here is usually worthwhile.

There is an important distinction between a Condition and a Warranty in law. This is worth noting and managing with care.

Special (or Supplementary) Conditions: Special Conditions of a contract are an extension and refinement of the General Conditions. They are used to provide additional terms and conditions, or to address the specific conditions that are unique to the deal in hand. Examples of the sort of thing that may be included in Special Conditions of a contract include the following:

- Special payment terms that reflect unusual conditions
- Ways of dealing with special risks
- Non-standard holdback provisions on payments
- Special needs for warranty and other guarantees of performance
- Insurance requirements
- Special needs for performance and other bonds.

Special Conditions are normally secondary to the General Conditions of the contract in the event of any discrepancy between the two.

Definitions: It is important to have a degree of clarity when setting out the terms of an Agreement. To this end, the definitions section sets out the specific meaning of certain words in the contract. For example, the word "Agreement" may be explicitly defined as the primary document that constitutes the entire agreement between the parties and supersedes any previous discussions, offers or other negotiations. It may be further defined as being the whole Agreement except when the Agreement explicitly refers to additional documents that are explicitly included. You will see that I have capitalized certain words throughout the book. These words have specific meaning and are explained in the Terms You Should Know section in the front material. It is normal practice to capitalize the words that are explicitly defined in a contract (for example, General Conditions, Special Conditions). This way it is easy to see that the words have special defined meanings.

Specifications: The specifications are the part of a contract that describes, in detail appropriate to what is being acquired, the requirements of the buyer.

Drawings: These are graphic representations of what is to be provided. They normally support and supplement the technical specifications. Typically, drawings are considered secondary to the specifications. If there should be any difference between the two, the specifications normally govern.

Appendices: Appendices are used to add background information. In a construction project, this information may include local bylaws, geotechnical reports or as-built drawings of existing facilities. For a systems project, details of data structures, hardware at the site and specifications for interfacing with other legacy systems may be included.

Addenda: It is not unusual for details to change during a contract bidding process or during negotiations. Such changes are sometimes documented through the issue of an addendum or addenda (plural) to the original contract.

In an appropriate combination to suit the deal, these documents form the basis of the contract documentation. They should describe the obligations and benefits for each party with sufficient clarity that both participants in the exchange understand the same thing. In addition to the terms set out in the contract, we need to consider some other factors that affect the way the parties are expected to behave and interact.

There are a number of components that may be used in the documentation for a contract. Together they should define the exchange of value between the contracting parties.

1.1.6 Obligations Outside the Terms of a Contract (i.e., Legislation and Tort)

In addition to the requirements set out in the contract documents, other factors will affect how a contract really works. The two groups of factors that we need to consider are legislation (laws made by the politicians at the national or local level) and the law of tort.

There is more to a contract than just the contract itself: You cannot contract outside the law.

Legislation creates laws and bylaws. It is useful to be aware of these. For example, in many North American jurisdictions, there is an Act (usually at the State or Provincial level) that protects suppliers from non-payment. This is variously referred to as the Lien Act, Mechanics Lien Act or Builders' Lien Act. These pieces of legislated law cannot normally be contracted out of, and in some areas, the law is written so that any contract that is covered by the act is deemed to be amended to comply with it. A common exception to this is work completed for federal government agencies. Once again, check with legal experts!

Legislation needs to be checked in the jurisdiction in which a contract is consummated and managed. (Sometimes a contract sets out which jurisdiction will be used in the event of a dispute).

The law of tort was mentioned earlier. This precedent-based law can be used as a basis for resolution of a dispute or for recovery of damages. You do not have to have a contract with the person who damaged you to entitle you to pursue a civil claim. If, however, you are damaged under the terms of a contract, as the injured party you may well have a choice to pursue your claim either under the terms of the contract or in tort. Usually your lawyer will advise you to pursue the most advantageous course for you. Again, check that the rules for engagement, particularly with government agencies, do not amend or waive the normal rights and obligations in law.

You can always sue someone in tort. If you are damaged under a contract, you may choose to look for recourse under the terms of the contract or in tort.

Sections 1.1.1 through 1.1.5 really do not do justice to the complexities of the law or to the subtle – and not-so-subtle – differences between legal jurisdictions. They serve merely to sensitize us to the main issues we need to consider or be aware of. Doing justice is also worth a quick and cynical look.

"Justice" and "Fair and Reasonable" are not the same!

1.1.7 "Legal" Versus "Ethical" or "Fair and Reasonable"

The law is rather mechanical and cold-blooded. Whenever we seek recourse in law, we are placed in a position in which we will either win or lose a lawsuit. The legal system operates in the public domain, so anything that happens in a courtroom becomes public knowledge. Not everyone wants that!

If you do not want to air your dirty laundry in public, stay out of the courts.

One observation about litigation (suing someone) is that it is a bit like fighting over a cow. One party is pulling at the horns and the other is pulling at the tail while the lawyers are busy milking it! Litigation is quite a popular sport in the world of contracting. In North America, construction contract disputes are a frequent reason for civil lawsuits, being second only to the ever-popular personal injury cases. Another little observation about litigation is the huge amount of time and cost that is required for a full-blown lawsuit (never mind the anguish and high blood pressure that goes with the process). Later on we will look at other ways to resolve and avoid disputes. These are addressed in Chapter 10.

The lawyers are more likely to win a lawsuit than you are.

The litigation process is all about credibility and application of the law. For credibility, it is necessary to have a stronger case with better evidence and witnesses than the other side. It is a bit of a game. It is possible for the side with the best lawyers, expert witnesses and supporting documentation to win a case, even if – arguably – they should have lost. So much for credibility. After all, what does a judge do except judge? Just about the most important thing that a judge can assess is the credibility of the witnesses and the evidence. Now let us look at the law.

Judges judge. They base their judgement on observation of your credibility and on the law. Your credibility is critical.

Justice is portrayed as a woman with a blindfold, holding the scales of justice and a double-edged sword. The symbolism is important. The blindfold tells us that justice is (theoretically) blind to who you are and your station in life. Reality, however, is that justice cannot be blind to the quality of the lawyers and other experts that a litigant can afford to retain. The scales of justice weigh both sides and tip in favor of right. The double-edged sword reminds us that the impartiality of the law can work both for and against you. You may think you have a case and, as a result, you pursue it through the courts only to find that the law interprets things differently, and you are at the losing end of the argument. If we think about it, this happens half the time. Anyone who goes to court does so because they believe they are right. Half the people who do this come out at the wrong end of the decision, and so lose when they thought they would win.

Justice is not about being fair. It is about being legally more right than the other party.

The law is not always clear. If it were, we would resolve our differences more easily without using legal recourse. Courts do not interpret contract terms if this can be avoided. They simply apply the law. "Not fair" has nothing to do with the law. That is the commercial deal you entered into, and the courts normally assume that you negotiated with the other party on an equal basis. The assumption is that the deal was struck on a level playing field. The fact that many people in the business of contracting do not feel that they are always able to negotiate with equal power or authority is neither here nor there!

The courts are normally concerned with the law, not with fairness or ethics.

For an example to illustrate the last two points, see Have I got a Deal for You.

Have I Got a Deal for You

A few years ago, I was involved as a partner in a software company. The company was funded by a couple of venture capital funds. It had produced a product that almost worked, and it was in the process of seeking the next round of funding to allow us to complete the product and to launch the marketing program in earnest. Our principal investor at that point had a clause in its funding contract that allowed it to buy shares at any price that any subsequent investor purchased our shares for, if the price was less than that paid by the principal.

At this point, we had successfully sold our product to a large multinational corporation. Another company was sniffing at our door because we had developed some technology that was of considerable interest to them. Our principal investor saw an opportunity to sell the company and make a profit on their investment.

As we looked for the next round of funding, we found it increasingly difficult to get any interest. We found out later that the investor network was riddled with rumors of our failure to produce a product and our imminent demise. We were desperate to ensure payment to our staff and did not have the money to do so. There was a deal on the table to purchase our company by the company that had expressed interest in part of our technology. Our principal investor told us that the fund they had used so far could not invest, but another of their funds could. The problem was that if we wanted to pay our staff, we would have to make a quick deal that would be at 1/10 of the price of shares that the first fund had paid. They had read us well. We agreed and instantly diluted our shares by increasing their interest ten-fold. This put the original founders in a minority share position and the venture capital company in control. They sold the company for a profit – on paper.

(Cont.)

Have I Got a Deal for You (cont.)

They had legally taken over the company and wiped us out. We learned later that the rumors about our failure tended to lead back to this same venture company that took us over in this legal but questionably moral way. Payment for the company was based on certain performance criteria that could only be met if specific key members of the team stayed with the new owners. Knowing this, one of the key players left. The performance criteria were never met, and the venture capital company did not get paid in the end. Their legal win turned into a real loss. But, perhaps, in the end some justice was done.

1.1.8 Contract Failure and Termination

Contracts can fail for reasons other than breach. Section 1.1.8 provides a quick review of the more important ways that a contract ends up not being one in the first place.

Mistake

A mistake can be made in the formation of a contract. One party may be thinking it is acquiring one thing while another party thinks it has sold something different. The wording in the contract may be sufficiently fluid or ambiguous that the mistake is not identified until later.

Consider the following. I want to sell a shipment of cotton clothing. It is all made of brightly colored material and is in a warehouse outside the city. The shipment is in cardboard boxes labelled "Corona Clothing". You go to look at this shipment and what you see is, in fact, a different shipment that is also in the warehouse. It happens to be made of better quality brightly colored cotton. I did not know about this other shipment in the warehouse. It, too, is in boxes labelled "Corona Clothing". You see this second shipment and agree to buy it at my asking price (based on my – different – shipment). You are unaware that there is another set of boxes in the warehouse because you stopped as soon as you saw what fitted my description of what I was selling. Here a genuine mistake has been made and the contract we entered into would be annulled.

When a genuine mistake is made in the formation of a contract, the contract is not binding. Beware, though, because this may not be true if only one party makes a mistake (see Just Kidding).

Just Kidding

A dam was to be constructed. There were several soils investigation reports available. One of the earlier ones had not identified a condition that would cost a significant amount to address based on the existing design of the dam. Subsequent investigations of the site revealed and verified this situation. These subsequent investigation reports were suppressed and only the original report was included in the material made available to the bidders for the construction of the dam. The bidders were, however, asked to complete their own soil investigations.

The successful bidder – who was awarded the contract – relied on the owner's reports included with the bid documents. The contractor subsequently expended a considerable amount of time and money on dealing with the situation. This additional time and money was not included in the bid, because the soils conditions were unknown to them at the time they prepared their proposal to do the work. They claimed compensation from the owner. The claim was rejected; the owner relied in part on the requirement for the contractor to complete its own investigation before bidding.

In the court case that followed, the other soils reports identifying the problem were used by the contractor to suggest that the existing conditions were known and were effectively hidden to misrepresent the real situation in order to gain unfair advantage at the expense of the contractor. The contractor was awarded damages.

In a similar case in which soils reports showing unfavorable conditions were suppressed, the contractor was also awarded damages as a result of the owner's misrepresentation which had led to frustration of the contractor or to recovery of costs associated with a benefit that the client would have obtained if there had been full disclosure of known facts at the outset.

Misrepresentation or Fraud

Misrepresentation or fraud differs from mistake in that someone is deliberately trying to represent that the deal is better for the other side than it really is. Someone is deliberately misrepresenting a situation to gain a commercial advantage under the contract. In such a situation, the injured party has rights for redress.

My own experience in another situation mirrors just this type of misrepresentation many times over. In one case, on a construction project, existing problems on a site were known by the client and were not disclosed to the bidders. The project ran into trouble because of these conditions and we offered the client a solution – even suggesting that the cost be shared between us. The client threw us off the project for not covering the full cost. With hot tempers, this led to a lawsuit that was heard in court some five years later. The original solution would have cost the client half of about $40,000. In the end, the client lost its case and had to pay its own costs plus ours. We were awarded $1.5 million in legal and other costs. By the time we had paid all of our experts, lawyers and subcontractors what we owed them, we were about $10,000 out of pocket.

Misrepresentation of facts as a basis of a contract leads to frustration.

Frustration (or Impossibility)

Frustration of a contract is the outcome when one party cannot perform its obligations as a result of a situation that was effectively not contemplated by the terms and conditions of the Agreement. It typically applies when an implied condition of the contract does not occur or ceases to exist without the fault of either party.

Frustration of a contract occurs when one party to the contract cannot perform its obligations for pre-existing and unknown reasons at the time of contracting or by virtue of a change in circumstances not contemplated by the terms of the contract.

1.1.9 Recourse in Law

In common law jurisdictions, the injured party under a contractual dispute usually has an option to pursue its claim in one of two ways. The first is under contract law, citing the contract and the failure of the other party to meet its obligations. The other option is to work on the basis that the contract effectively does not exist and the claim is made in tort.

If you are the injured party in a contract, you can sue in tort or under the terms of the contract. In some cases, both options may be open.

Before either option is considered, a series of steps need to be taken by the party that feels it has been injured. To make a reasonable claim, we need to determine what damages we are trying to recover. Then we need to quantify that damage and justify the basis on which we have arrived at that figure. Usually we reduce a claim to a request (I use the word loosely!) for more money. We need then to determine the basis for the claim and find support for it in the contract. If the claim cannot be supported by the terms and conditions of the contract, then we need to consider the possibility that there is no entitlement.

Before claiming for damages under a contract, make sure it is worth the trouble and that you are entitled to such a claim.

1.2 Changes that Influence Contracts

In the second part of Chapter 1, we consider the changes over the past few decades that affect contracts and the business of doing business with others under a contract, purchase order or other agreement that is legally binding. We will look at the following topics.

- Change in the past 30 years – contracting, technology, business, social (including politics, management styles, language and wealth distribution)
- Effects of these changes on how we do business
- Effects of these changes on contracting options
- What has been found by CII (Construction Industry Institute) and CRINE (Cost Reduction Initiative for the New Era)
- Innovations in industry contracting practices

The world is changing faster than the contracts we use.

1.2.1 Change over the Past 30 Years

Technology has changed our way of doing business in a myriad of ways over the past 30 years. Yet, in that time, we have moved very little in the way we do business with contracts except in the most innovative organizations or in isolated one-time situations. Or have we?

First let us look at a few of the technological changes that have affected us in the world of contracts:

- Word processing has allowed us to replicate contract terms and conditions and technical specifications at the push of a button.
- Computer-aided design tools have allowed us to re-use design elements – and sometimes entire designs – with little effort other than changing the name of the client.
- Telecommunications have facilitated the effectiveness of distributed project teams.
- The Internet has opened doors to new ways of doing business.
- The time-to-market imperative has forced us into different ways of accelerating product delivery, whether that product is a new cell phone, a factory, an office building or a new toy that needs to be in the stores in time for Christmas.

There are countless other technology shifts and changes that affect us and that we do not even give a second thought to. Some of these technological advances include Velcro™, the cell phone, fax machines, yellow Post-It Notes™, and Rollerblades™. Maybe some of these do not affect our contracts directly, but many certainly affect the way we do business, record events, determine reality and communicate with each other. These all affect the way we do business, so they must affect our contracts when business is between two or more legal entities (except, maybe, Rollerblades!)

There are many changes that affect the way we do business. In turn, they affect any contracts that are implicated in the business process.

1.2.2 Effect of These Changes on How We Do Business

The changes identified in Section 1.2.1 and many others have had a profound effect on how we do business. Some of these effects are apparently contradictory. Here is a partial list to get us thinking.

- We now have a global economy.
- Knowledge that was considered proprietary a few scant years ago is now available on the Internet.
- The half-life of technology is still decreasing.
- The pressure to deliver faster continues to increase.
- We are becoming more specialized than ever, regardless of the business we are in.
- We need to rely more on external expert suppliers, consultants and contractors than ever before.
- The process of delivering a product or project is more complex than ever before because it involves more regulations, stakeholders, suppliers and third parties than it did previously.
- To stay competitive, we need to find cost and time efficiencies or other innovations that we did not even know existed before and yet still produce a quality product.

There is a myriad of change in terms of what affects contracts and how it has altered over the past 30 years.

Let us pick the last item on this incomplete list. What are these corners we need to cut, and why do they exist? One of these corners is the learning curve that exists each time we find a new supplier of goods or services. They need to be "trained" to meet our needs. This training can relate to understanding how we do business, what we need in order to pay the supplier, what our tolerance for quality fluctuations may be and how receptive we are to innovation or change or to ideas not invented here. The equivalent exists also for the supplier. One way to reduce the learning curve and to build loyalty that should serve us well in a seller's market (if we are a buyer) or a buyer's market (if you are a vendor) is an evergreen contract or an alliance. These long-term arrangements mean we do not have to re-train each other for each event, contract, project or crisis! Cutting this corner creates a new problem: How do we know that the contractor or supplier is going to continue to be competitive if they feel they have a captive market?

Some of our solutions to issues that are developed by changes create new problems for us.

Let's try another corner: Actually having a contract in place to do the work before we start. An increasing amount of work is being done in the absence of a current and active contract. There are two primary areas in which we frequently deliver goods and services without any formal contract. One is when a letter of intent has been issued but no formal contract has yet been executed. The other is when work is done that will later be covered by a change order but which is currently outside the terms of an existing contract. Often this work is done in good faith, particularly when there is a clause in the principal contract that states that the contractor is not entitled to any payment for any such work started before an agreed change order is issued.

So, what does this tell us? We are doing work on the basis of relationships with people rather than on the letter of the contract. We are moving more in that direction as pressure for early delivery, the need to collaborate and share technologies and other business drivers require such a shift.

Even when we do not change the contract itself (as a legal document), we are modifying our behaviors regarding how we deal with each other. This can effectively change the terms of the contract.

1.2.3 Effect of Changes on Contracting Options

It seems that everyone wants everything faster than ever before. Time to market is of huge importance when we live in a world where there are lots of great ideas and opportunities for the quickest amongst us. In addition to this, we do not like to see our investments not earning money for us. So once we start to spend on a project, we like to see it finished as quickly as possible. This principle extends to the public sector. Try closing a bridge for a few months for refurbishment and you will quickly discover the political fallout and claims of lost business from people on both sides of that bridge! The most conventional solution to this type of situation is that we overlap design and implementation. We do this rather than wait for a complete design before tendering the rest of the work based on complete technical specifications and associated documentation – such as drawings, commercial terms, information about existing conditions and so on. In fast-tracking like this, however, we buy a whole set of new problems that are all too often underestimated or even ignored.

 Time to market has pushed the limit on conventional contracting and forced us into fast-tracking.

Another manifestation of the need to attack the time-to-market challenge (or necessity) is seen in the manufacturing business, as well as in new product development. It is not unusual for the client to do much of its own design, subcontracting only those parts that are non-competitive. What is defined as non-competitive has also changed over time, as we need the expertise of our suppliers more than ever before. This "manufacturing" mindset has translated into some interesting initiatives in the construction sector, for example. Here, there is a growing awareness of the obvious. The people who manufacture pumps probably know more about manufacturing them – and maintaining and operating them in different situations – than either the people who buy them (and operate them in only one or a few specific situations) or those who specify them (but do not install, maintain or operate them). *Quel surprise!* Now the manufacturers are, albeit occasionally, being asked to have input to the specification of the pumps, motors, liners and other components. And not just the pump manufacturers, but also those who manufacture other items used in construction. Through constructability reviews, we are now seeing construction contractors applying their expertise to designs.

 Cost pressure has pushed us into degrees of collaboration that many people did not consider before.

This collaboration is enhanced and even challenged by the Internet. Let me explain. In the last few months of 2001, I came across a significant – and growing – number of companies that are establishing themselves on the Internet, offering services and deals that were previously too difficult to offer. These range from bulletin board services (with a bit of clever stuff behind them) that act as clearinghouses for project information to others that offer brokerage services for suppliers, designers, specifiers and buyers of goods and services. I have even come across a company that is offering free engineering of relatively standard products because they can broker and buy components more effectively. The company packages the "free" engineering with the installed product, making a profit through brokering and discount purchases that are not available to the regular buyer who does not purchase as much as the engineering company does. Simply using the Internet to transfer information and build relationships has created new and interesting synergies between companies that did not previously consider themselves as partners.

 The Internet has created opportunities for collaboration at a whole new level.

Selling goods and services on the Internet has generated new ways of doing business. We are repackaging what we offer our customers to provide competitive advantage to them through value and to us through more effective use of the resources we have to offer. There are many examples of this:

- We design in one country, produce drawings in another and manufacture products in a third.
- We develop software 24 hours a day by breaking the work into three stages and separating the work centers by about 8 hours. So we might specify and integrate in Montreal, Canada, write code in Delhi, India, test the work in Warsaw, Poland. Finally, we will return it to Montreal for integration. Protecting the intellectual property becomes a challenge – but this has been solved in some interesting ways.
- Some new products are developed in close collaboration with suppliers and contract manufacturers. They provide input to the design of mouldings, the mounting of equipment, the development of each other's products, co-ownership of manufacturing equipment and more.

Put another way, all organizations are becoming integrators of the technologies of others. Some organizations are recognizing that this involves a much closer relationships with suppliers and other business partners. A critical part of what we offer is a good working relationship with our clients and with our supply chain.

Competitive pressures are forcing us to repackage what we offer.

It is not unusual for businesses to add value in ways that are not traditional to the industry. One increasingly common opportunity lies in financing the work. This opportunity will increase as many traditional buyers (such as government agencies and some large corporations) struggle with finding the cash flow to support their need for spending. A good example in the Western world is the rapidly increasing infrastructure debt. This debt is what we need to spend on maintaining or repairing our infrastructure. The bridges, water supply systems, sewer systems, power distribution systems, roads, buildings and other artifacts that support our way of life are in growing need of attention. This attention needs to be in the form of replacement, expansion, refurbishment and maintenance that is overdue. One estimate of the infrastructure debt in the United States and Canada, at the end of the year 2000, was more than $450 billion.

This debt and the need in industry for retooling or development of new products and services are leading to opportunities to provide the required contracting services *and* the funding to support this. The repayment of the funds is often the result of some creative thinking. Better heating and ventilating systems are offered in exchange for a percentage of the savings in operating costs. Roads are built in exchange for toll revenues. Water treatment plants are built and operated privately in exchange for user revenues. New products are built and tested in exchange for shared revenues, stock options or other benefits.

The demands of attractive balance sheets and the funding shortages by traditional owners have created opportunities for more integrated packaging of projects and the services associated with them.

When we combine the need for faster delivery and the advantages of good working relationships with our client and our supply chain, we generate the ability to learn to work together. This in itself saves time, as we do not need to train each other in how we do business. We are, if you like, pre-tuned to the other party's way of operating, so we can respond faster and more accurately. This type of relationship is being formalized more frequently through alliances and strategic partnerships.

 Elimination of learning curves and familiarity with client requirements and other long-term needs are being addressed through longer-term business relationships.

There are many wild cards in the way the future will evolve. Here is one that seems to be of some significance. The human element in how we relate to others and the anthropological view of business relationships are starting to give us some interesting pointers to where our future may lie. Anthropologists are wonderful observers. This training and the associated knowledge bring insights that others miss.

As a keen amateur in this field, and with some active research interests that demand greater understanding of the human condition in a business context, I offer one observation in this arena. One big change in how we do business will grow out of a better understanding by more business people of the nature of human interrelationships. As one example of how things may evolve, we will look at the role of trust in contracting relationships in more detail in Chapter 7.

 Studies into human behavior are beginning to make us challenge the way we think about contracting processes.

All of these elements, and other changes to the assumptions that have held true or remained unchallenged for so long, are beginning to influence the more adventurous people in the business of buying or selling goods and services. Some of the new ideas I have suggested turn out to be old ones when people tell me, "Now wait a minute, we've been doing that all along". Here are some examples:

- How about a contract that is based on trust?
  "We issue change orders after the work is done. Under our contracts, the contractor is not entitled to any money if they start the work before we issue a contract. They (and we) are working on trust all the time."

- Here's an idea – why don't we do away with change orders altogether?
  "We do business in China. There, the basis of doing business is that a contract is the starting point of negotiation, not the end point. Everything is negotiated, even what we think is clear in the deal!"

- OK, why go for the cheapest bid if it just offers the lowest initial price, not the best value?
  "On the last project, we paid the bidders to come up with a design concept, execution plan and matching budgets, options and any other ideas they might have to make the project a success. We picked the best package rather than the lowest apparent bid. Mind you, we had to bend the company standards to do this!"

Old ideas about what constituted a good contract are being challenged on a daily basis.

*It's written in ancient Egyptian hieroglyphics -
so we know nobody has read it.*

1.2.4 What Has Been Found by the Construction Industry Institute and the Cost Reduction Initiative for the New Era

The Construction Industry Institute (CII), based in Austin, Texas, was a development or outgrowth of the Business Roundtable, which had produced a report in 1983 entitled "More Construction for the Money". This was the summary report of the Construction Industry Cost Effectiveness Project. The CII is made up of a group of organizations that have an interest in making construction more cost-effective. Many of the members are involved in the petrochemical industry – one that is heavily reliant on construction as part of its business process. Some of the studies undertaken by the CII have included ones related to contracts and the issues surrounding them.

One specific study led to the development of a simple software tool that provides a guide to the "cholesterol level" of a project for which data is entered. This cholesterol test provides an indication of the probability of ending up in a contractual dispute on the project. The questions asked include ones that establish whether the two parties have done business before, what the relationship had been like and how complete the design was at the time of tender. The higher the cholesterol count, the higher the probability of a dispute. The user can then review the likely causes and will be able to develop a better contracting strategy to reduce this risk. What is interesting is that the study behind this product clearly identified indicators of potential future problems. The resulting product is predicated on the fact that contracting problems are predictable.

 The Construction Industry Institute found that contracting problems are somewhat predictable.

Another study by the CII identified what they labelled as the Continental Divide. This is the point at which a dispute moves from being resolvable to being a serious issue. This point – at the risk of oversimplifying the results of this study – occurs when people take on a position rather than try to find a solution. Interestingly, this mirrors the work of Fischer and Ury, made accessible through their book "Getting to Yes". Positional arguments are so invidious that I use them in a workshop exercise to illustrate how easy it is to get caught up in them. Once in such a position, we tend to fight to win. This implies the other side must lose. Guess what?

The other side needs to win too. Stalemate. Both sides lose and sometimes a lawyer or two (or more?) wins!

 Once a disagreement gets to a certain point, the opportunity to resolve it intelligently diminishes considerably.

In another interesting study, the CII investigated the potential for a correlation between the cost of projects and the level of trust between the parties to the contract. They found there was a statistically significant correlation. As trust increases, so the cost of the project decreases. The study suggested that there was a practical limit to the level of trust that could be applied. Beyond this was what the report referred to as "blind trust".

This is a significant finding, especially because much of what we see in contracting practice is based on low trust behaviors by all parties. This has led to a new study – part of my own research program – that looks at this phenomenon in more detail.

 There is a relationship between the cost of a project and trust between the contracting parties.

 The greater the trust, the better the value obtained for the expenditure.

 There is a practical limit beyond which we cannot take trust.

Accessing trust has more to do with the relationship between people than with the contract that is negotiated and signed. This said, the contract is the starting point of the business relationship. If previous relationships existed, then the contract is a transition point from one experience to the next.

The wording of many contracts is designed to establish the buyer's control of a situation to which they have committed to spending money. The buyer needs to have some assurance that they are getting value for what they spend. The language that is generated by this concern and by the many bits of advice that we hear – such as "Buyer beware" and "A fool and his money are soon parted" – has a tendency to be aggressive. This aggressive language gets in the way of building trust between parties and can even erode trust that was built on previous occasions. To address this first barrier, we need to reconsider how we structure our contracts. Wording is just a start.

To harness trust and the benefits we can derive from it, we need to develop more innovative contracts and contracting processes.

To survive, we need to evolve. Evolution in contracting has been driven in part by the changes we briefly looked at earlier in this chapter. The Cost Reduction Initiative for the New Era (CRINE) initiated by the British government has also been a mine of innovation. The driver for this initiative was waning royalty payments resulting from oil and gas exploitation in the North Sea. The cost of developing new fields was prohibitive. Smaller discoveries of minerals below the seabed were too costly to recover using current approaches. British Petroleum (BP) led the way with a number of pioneering initiatives. These are more fully described in "No Business as Usual" (Knot 1996). The concept was simple: Put the contractor on the same side of the table as the buyer. By creating win-win opportunities, the very nature of business was changed and phenomenal savings were achieved in time and money on a number of projects. A parallel initiative in Norway led to comparable advances in competitiveness. The sustainability of this approach may seem in question today, as it does not appear to have spread as widely as the results would suggest it should.

Change is forcing innovation on contracting approaches.

Some of the changes that we see in contracting processes are re-casting, re-packaging or re-combinations of old practices. Fast-tracking is not new. Combining financing with engineering and construction is not new. Incentive schemes are not new. Supply chain management is not new. Long-term contracts (evergreen contracts) are not new. Repeat business (strategic alliances) is not new. What is perhaps new is the spirit behind what is going on, a spirit based on greater trust and reliance on each other. Now, wait a minute, the concept of professional or trade guilds, handshake contracts and doing business with people you know is not new either . . .

There is probably nothing new under the sun. Most of the changes we see are modifications or adaptations of practices that have been in existence for a long time, but may have fallen into disuse.

So if nothing is really new, why do we need to innovate? Why are so many people continuing to test ideas and different ways of contracting? The answer lies in the relevance of what we do to today's marketplace, to the challenges we face in getting our business done competitively and to our best advantage. The traditional approaches to contracting were based on traditional approaches to business. The need for speed, new technologies, increased regulation, greater (third-party) stakeholder involvement and other changes are forcing us to try other approaches to contracting that help us achieve our objectives in a more complex and uncertain world.

To understand the need for change and innovation in contracting, we need to understand what is making more traditional approaches less valid in today's business environment.

1.3 What has Changed

A number of things have changed that, combined with the way we contract, create many of the problems we see today. In Section 1.3, we will look at issues that have emerged because of the changes we considered in Section 1.2.

Some of the more important things we touch on in Chapter 1 are revisited in more detail in Chapters 2 to 10. For now, we are interested in setting the scene for considering changing how we do business with other organizations.

A fascinating study by John Kamara shows that we are not too good at understanding our client's needs. This study suggests that we are likely to impose our solutions rather than really hear what the client's problem is. In other instances, we have a habit, as a client, of offering our designers and contractors a problem in the form of our own solution. This may well be the right solution, but it might not! The way we specify what we are contracting to purchase often limits the opportunities for creativity or transfer of technology or even competitive advantage.

Not taking advantage of suppliers' knowledge can limit our ability to be innovative and to offer or gain competitive advantages. These are lost opportunities.

Customer expectations are not being met. This is a bold statement and clearly is not always true. But three private studies I conducted or was involved in for consulting engineering firms have consistently revealed the following interesting results.

- Customer satisfaction is NOT related to delivery of projects on time, within budget and to the original specifications.

- Customer satisfaction IS directly linked to the ability of the consultant to truly understand the needs and expectations of the client.

- Repeat business is generated based on relationships with people.

- Technology offers an advantage only where there is a specific need for that technology or where it serves to differentiate in a competitive situation.

- Price rather than value for money wins when competitive bidding is the rule.

- Value, rather than price, rules in the selection of a consultant when a relationship exists and the client is aware of the skills offered by the people assigned to the project.

We need to work harder at understanding and delivering to customer expectations.

Incentive schemes that go awry are not uncommon. It seems to be a fine line that separates an effective contractor incentive scheme and a problem in contract administration. When incentive schemes become a priority for either the contractor or the client, then managing that part of the contract can dominate and even distort the rest of what the contract is about. In such situations, we often see a greater propensity for disputes and lower levels of trust between the parties involved in the deal.

 Although incentive schemes may be useful in contracting, there is a significant risk that they (rather than the job in hand) will become the primary focus.

Delays and their impact are often not fully understood by all of the participants in a contract. For example, a delay of one day that pushes work over a longer break (such as Christmas and the New Year) will result in a longer delay. Also, if a specialist subcontractor is delayed, it may not be available again for a considerable amount of time, and the project will be delayed until the subcontractor is free from other commitments. Also, delays – in themselves – disrupt the planned flow of work and have a ripple effect that is hard to identify and evaluate. Just part of this effect is the impact of delays on team morale. Are we saying it is acceptable to slow the work down? Are we turning our backs on the effort to date to maintain a challenging schedule?

The cost of changes to the contract – in the form of technical or scope or scheduling adjustments – is routinely underestimated. Changes are often the cause of delays (see Section 1.2.2). In addition, the cost of a change will vary depending on when the change occurs (see What Does a Change Cost?)

What Does a Change Cost?

This project included some sophisticated electronic controls. A remote sensing, control and data acquisition system was to be included and it was to be integrated with other corporate systems. The client owned and operated a number of oil and gas pipelines and wanted to be able to more accurately track product for its clients. It was also interested in more effective client billing, as well as better environmental audit practices. Given the timing of the project and the state of current technology in the subsidiaries it was trying to integrate, this was a challenge.

Uncertainty about the scope of this project led to a request for a separate price for inclusion of some additional features that would allow the client organization to add specific components at will, if the budget and schedule permitted. The contractor for the work was asked for this information at the time of tender, and submitted a price based on making that change at the outset. The client made the change as the project neared completion, and the contractor argued that the separate price did not reflect a change at this late stage. Just some of the differences they listed included the following items that would increase the cost of making the change.

- Remote sensing devices were based on the old scope. These had now been installed and would have to be removed, new ones ordered and these new ones installed. There was a restocking cost associated with some of the units that were now redundant and others would have to be scrapped. The original price was for ordering the upgraded units and installing them in the first place.
- Programming was complete – the system was being tested. Much of this programming would have to be redone. In particular, programming of interfaces with existing client systems would be impacted, adding months of delay and many hundreds of hours of rework. The original price included only for programming based on the extended data strings that would have been needed should the change have been made at the outset.

(Cont.)

What Does a Change Cost? (cont.)

- The addition of three new sets of data in the signals coming from site and in the system that would be handling it was a profound change at this stage, as the database would need redesign, legacy data would have to be reconverted and testing would need to be redone.

Clearly, timing is everything when it comes to changes!

Design changes and their impact after the contract is awarded are clearly manifested in What Does a Change Cost? In other cases, these changes emerge as "design clarifications". A common argument is that the contractor, being experienced and knowledgeable, should have understood the real intent of the designer or buyer even if it was not actually stated! Contractors are not mind-readers. Designers are not perfect and Owners sometimes expect perfection. This combination of factors can lead to disputes and possibly worse.

Risk assignment and associated misunderstandings are another big cause of disputes in contracts. It is quite common for contracts to be written so that the authoring party (usually the buyer) is substantially divested of risk. The fond belief is that the contractor will take this risk without asking for any compensation. This belief seems to be stronger when the buyer tenders and selects the cheapest bidder based on a stipulated price contract. After all, to be competitive, the contractor cannot add lots of premium to the bid price and still win the work. The winner (and, again, I use the term loosely!) will be the contractor that has put the lowest (zero?) price on these risks or has ignored them completely. Perhaps more alarming is the apparent agreement by many contractors that they cannot afford to add a premium to the price to cover risks. But they do. If they did not, they would be out of business.

The real cost of exculpatory (weasel) clauses was hard to identify. We will look at this in more detail in Chapter 6. For now, let us just bear in mind the range of costs associated with the five most common exculpatory clauses used in the construction industry in North America. These clauses address the cost of delays, existing site conditions, the quality of the design, compliance with regulatory and other requirements and liability to third parties. These five clauses cost the buyer between 8% and 20% of the total construction cost, depending primarily on market conditions. The value of these clauses is another matter. What does the buyer get for this premium? The answer is very little. We will look at this further in Chapter 6 and 12. For now, the point is that we need to re-evaluate our contracting habits and traditions and consider what is going on. There are real opportunities to reduce cost, improve performance and leave everyone better off in the end.

Procurement is a vital part of most enterprises' competi-tiveness and success. There are opportunities we cannot afford to ignore in the management of our suppliers as well as our customers.

Chapter 1 provides an introduction to some of the broad legal principles that apply to contracts in many countries, but specifically in the United States, Canada, United Kingdom, Australia, New Zealand and other countries whose legal systems are based on Common Law.

We use that one to lie to the boss.

Chapter 2

THOU SHALT NOT MIX UP THE WRONG WORK PACKAGES

> *Packaging Work: Combining the work and materials supply for optimal performance.*

In Chapter 2, we look at what we are contracting for. Essentially, we contract for goods and services. We may do so in a myriad of ways. The simplest and most "traditional" way is to contract for all goods and services required to complete a project using a competitively tendered stipulated price contract. As we move away from this relatively simple option to more complex contracting forms, we change the type of contract, how it is awarded, and what the scope of the contract is. The different implications, advantages and disadvantages of the options are presented.

2.1 Contract Types

There are three basic forms of payment terms for contracts, though they lie on a continuum, with many flavors in between. We can put contracts out for prospective suppliers of goods and services to bid on them, or we can negotiate the terms. The scope of work covered by a contract can vary extensively from supply of one item to a fully integrated package including financing and operation of the end product. The way we set about selecting the type of contract we want as a buyer is an important part of the development of any project. Contracting strategies set the DNA of a project, as they influence power positions, payments, who has responsibility for what and much more.

Selecting the right type of contract sets the DNA for a project. We will live with the effect of this critical decision until after the contract and project have passed into history.

2.1.1 Stipulated Price

Perhaps the most commonly used type of contract and also the one most people feel they understand is the Stipulated Price Contract. The degree of understanding and the expectations around what such a contract delivers are often not consistent with reality. One common misconception is that the price will not change once the contract has been signed. Another example of the existence of misconceptions regarding this type of contract is the fact that it is often referred to as a Lump Sum contract. This latter type of contract is a specific form of a Stipulated Price contract in which the contractor is paid at the end, when it has performed all its obligations under the Agreement. For this reason, it is rarely used, as most contractors consider it too risky and expensive to finance. Most stipulated price contracts provide for regular progress payments based on completed work.

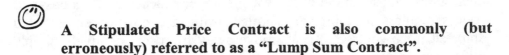

A Stipulated Price Contract is also commonly (but erroneously) referred to as a "Lump Sum Contract".

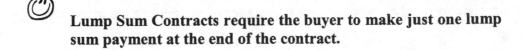

Lump Sum Contracts require the buyer to make just one lump sum payment at the end of the contract.

Variations of the stipulated price contract can be created through combinations of different types of scope (such as design/build or turnkey). Subtler variations include target costs, schedule performance clauses and other risk assignment options. As the potential need for additional management, coordination, or risk mitigation increases, so will the price for the work. At least the price for the work quoted by rational and aware contractors or suppliers will increase to reflect the added effort or risk associated with such a contract.

There are variations on these types of contract that may have a significant impact on the price and the risks associated with the contract.

Some of the most common variations on a stipulated price contract are worth a quick review. In the United States and Canada, most states and provinces have lien legislation in place. The lien act in each jurisdiction is unique, but a common feature is that anyone paying for an improvement is required to hold back a portion of the payment. This money is typically deemed to be held in trust for the suppliers of goods, labor or services to the contractor in the event that the contractor fails to pay them. The holdback is released after a specified period following a defined completion point for the work described by the contract. Deferral of payment is also used against deficiencies and disputed items. Incentive schemes and bonus-and-penalty clauses may also be added. Sometimes the same effect is achieved through the use of a liquidated damages clause. A common area covered by such a clause is schedule performance. In this sort of situation, the contract will stipulate an amount for the damages sustained per day or for any other period that the completion of the project is delayed.

 Variants on a Stipulated Price Contract include use of mobilization fees, holdbacks or payment deferrals and incentive schemes.

 Holdbacks may be a mandatory requirement in some jurisdictions.

Holdbacks and other payment deferral methods are used by owners as a foil against performance failure by the contractor, particularly towards the end of a project.

The variations we have looked at so far are generally ones that tend to penalize the vendor in the event that it fails to perform to the expectations of the buyer as articulated in the contract. Another variation of this type of contract is the inclusion of a payment upon award to cover the cost of mobilization. This is more common on international contracts or in situations in which highly specialized equipment or other components are required before the work – and therefore progress that can be paid for – can be started. In such a situation, the contractor would face a substantial financing cost and associated risk, as well as reduced capacity to do work, because of the significant amount of working capital tied up in the project. As most buyers have already obtained funding for the work, and may well have better financing than the contractor, it is often in the best interests of all concerned that financing of the up-front cost of mobilization is covered by the buyer.

Mobilization costs are paid to the contractor to reduce the contractor's financing charges at the start of the project.

When applying "voluntary" holdbacks or advances of payments to the contractor, consider who is paying for financing of this money.

Over the past decade, buyers have tried incentive schemes of various types with varying degrees of success to encourage better performance by the contractor. If time is critical, the owner may pay the contractor a bonus for every day that the completion date is brought forward. As long as the cost to the contractor is lower than the bonus being paid, the contractor may be tempted to try to accelerate the project. If the risk of failure is high, the incentive is less likely to be pursued. Other incentives may be used to encourage a low accident rate. If the adopted design and construction process permit, incentives to offer suggestions for reducing cost may be feasible.

Incentive schemes for contractors can be used to encourage specific aspects of performance.

A real risk associated with use of incentives is that the focus on the critical element thus created will defocus attention on other aspects of the project. For example, an opportunity to finish the project early by cutting corners in quality may prove too difficult to resist and will yield a more serious problem (such as a shutdown of the facility that was built so quickly) resulting from the compromised quality. Similarly, a heavy focus on cost may lead to unnecessary and even more costly delays. The cost of the delay may not appear as a part of the cost of procurement but may be in the form of a lost opportunity or reduced or delayed revenues.

Cost based incentives challenge the quality of the product. Time based incentives challenge safety, quality, and so on.

It was clever to specify pre-assembling everything, but how will we get it through this pre-assembled door?

Some safety-conscious buyers – particularly construction buyers in the petrochemical industry – pay contractors a bonus for low, or zero, lost-time accidents. This may well lead to better safety on the construction site, but it also tends to yield a measurable number of "walking wounded".

These people are recognizable on a construction site as they wander around with brooms sweeping or doing other light or nominal work so that they stay on the job. As long as they are on the job, they do not become a lost-time statistic! When this game has been identified, the next step is to measure all accidents that require medical attention. Once the medical attention is addressed through "first aid" on site, then another step in escalating the metrics will likely be taken.

Safety-based incentives can lead to creative reporting of accidents, injuries and other safety problems.

Common targets for incentive schemes are schedule acceleration, cost reduction and safety standards.

2.1.2 Unit Rate

The idea behind a Unit Rate Contract is that a price can be obtained for different types of products or services in which the final amounts for each cannot be precisely determined. This type of contract eliminates the risk for a contractor associated with assessing the wrong quantity that may be required by the buyer.

Unit Rate Contracts are particularly useful if the quantities of various components under the terms of the contract are unknown.

Unit rate contracts for construction are more common in the United Kingdom than in North America. In the United Kingdom, it is common for a buyer to provide a bill of quantities that is based on a standard method of measurement. Each contractor prices the same quantities, thus allowing the buyer to compare apples with apples in the contractor selection process. In North America, Unit Rate Contracts are priced on units that are not necessarily based on standard methods of measurement. Furthermore, each contractor bidding on the work may be required to measure the quantities associated with each unit. Either way, the Unit Rate Contract approach helps the owner compare potential costs based on assumptions regarding quantities rather than known amounts of work. It is quite common for engineering and other design services to be priced in this way, as the total number of hours to be expended on engineering normally cannot be determined on anything but the most common and repetitive products.

 The underlying principle of Unit Rate Contracts is that the contractor will be compensated for different types of work based on rates for that type of work.

When unit rate contracts are awarded, the objective is to be able to measure actual quantities and pay for them. A number of issues arise. Let us consider three common ones. First, buyers are concerned with obtaining value. So if engineering is paid for by the hour, the level of productivity comes into question. An invoice may be queried based on the perceived value. Hours spent on fixing design errors may be challenged. Rework time, training opportunities, and other uses of time for the people charging time or costs to the project need to be clarified. A second example is how items are measured. Consider digging a trench. Do you pay for what is actually dug or only what is efficient? The standard solution is to identify payment lines: If the contractor digs beyond those lines, then the owner will not pay for the additional excavation or the associated additional backfilling. Alternatively, a unit of length is used for the trench rather than the volume of material that is dug or backfilled. A third example is the border between one type of work and another. Consider the difference between excavating hard soil and excavating rock. If there are two different rates, then how is the boundary between them defined, especially if there is a transition from one class to the next? Here again there are "standard" answers. One such answer is to test the material. A practical test is to see if a ripper blade attached to a certain size of bulldozer can rip the material in a certain way (such as to a depth of 200 mm in a single pass).

 The description of a particular type of work needs to be sufficiently clear to allow all parties to understand and agree on what is involved.

To obtain value from the process, the documents on which the bidders' quote and ultimately the successful contractor's contract are based must include estimated quantities for each of the items for which a unit price is being requested. If this forms the basis of a contract, then often the contract will recognize that the quantities are approximate and that, as a result, the effect on profit and overhead (typically included in the unit rates) may be significant if these quantities are very different in reality.

Consider a trivial case of a contract with one unit rate and a quantity of 100,000 units at a price of $50 per unit. This gives a total estimated revenue to the contractor of $5 million. If the fixed costs for the project (mobilization, field set-up, supervision and other direct costs) are $1 million, profit is $200,000, overhead is $300,000 and the cost of producing each unit is $35, then the contract works if the quantities are equal to or larger than the assumed 100,000 units. If, however, only 70,000 units are required in the end, the payment to the contractor is:

	70,000 units x $50/unit	= $3,500,000
The cost is:	Fixed cost	= $1,000,000
	Overhead	= $ 300,000
	Production cost = 70,000x $35/unit	= $2,450,000
	Contractor cost subtotal	= $3,750,000
Then the contractor makes a LOSS of		= $ 250,000

To stop this sort of situation from happening, the pricing of unit rates may be renegotiable if the real quantities differ by more than a certain amount relative to the estimated ones. A common range used in this type of contract is an expected accuracy of +/- 15%.

Quantities of work to be done under each unit rate are normally identified in at least approximate terms. Variances that are significantly beyond a reasonable margin (often +/- 15%) of the estimates can lead to renegotiations of the unit rate for that type of work.

Unit rate contracts are common in some jurisdictions, as they are used with a bill of quantities (or bill of materials) to effectively award a stipulated price contract with some change provisions built in.

When bills of quantities are used as part of the contract, this often serves to save contractors from having to complete their own quantity assessments for each project they bid on.

2.1.3 Cost Plus

Cost Plus Contracts are used in situations in which time to market or rapid return on investment is of particular importance. In these situations, it is common to see a number of operations that traditionally and efficiently happen in sequence instead occur in parallel. This overlap in the normally sequential operations often leaves people involved in later operations incompletely informed about what they are required to do. In construction, design is partially complete when the contractor starts work on construction. Thus, the final cost of what is to be built cannot be accurately determined, even within the larger range typically accommodated by unit rate contracts. Cost Plus Contracts are commonly associated with fast-tracking a project.

The primary driver behind using Cost Plus Contracts is normally the need for speed. There simply is not enough time to put all of the pieces in place for a Stipulated Price or Unit Rate Contract.

Heavy reliance is placed on either the integrity of the contractor or the ability of the owner to audit the books of the contractor in a Cost Plus Contract. It is not uncommon under the terms of such a contract for the buyer to include some provision that allows it to review or even audit the contractor's books regarding the project.

Cost Plus Contracts require either a degree of trust or a detailed audit system.

There are several flavors of Cost Plus Contracts. All of them are based on a generalized model of the owner paying all the direct costs associated with the delivery of the goods and services required under the contract plus an additional amount. The differences lie in how this additional amount is measured and in how the "direct costs" are interpreted. In the case of the latter component, there is considerable confusion in the construction industry regarding what is included in the various components of costs in contracts.

Cost plus contracts are based on the cost of doing the work plus a specified payment to the contractor.

As for the "Plus" part of the contract payments, there are several models. One of the oldest is the use of a percentage to cover overhead and profit. This is probably a remnant from the roots of this type of contract in the Unit Rate model discussed in Section 2.1.2. Some design fees are still paid as a percentage of the final cost of a building. This is the classical way in which architects have been paid in the past. This format has the unfortunate effect of rewarding people who spend more of the buyer's money than is necessary to get the job done! As a result, it is falling into disuse.

Today, there is little use made of cost plus a percentage fee – this model of contract encourages cost increases by rewarding them.

Another model used to address the "Plus" component is a fixed fee that should cover both profit and overhead for the contractor. This serves to encourage the contractor to finish as efficiently as possible. Speed reduces the time that overhead is burned up in the contractor organization. Efficiency is encouraged (theoretically) by allowing the contractor to manage its staff so that as much of the fee is retained after costs. There is, however, a downside to this approach. The contractor can be efficient in the use of its staff at the cost of the project (higher direct costs paid by the Owner while the Contractor spends less on supervision and management). It may be that materials are not purchased as efficiently, labor usage is not as well planned, and many other sources of inefficiency in the spending of buyer money creep in. Balance is required to maintain overall effectiveness and a win-win situation for this type of contract to be truly effective.

Cost plus fixed fee offers a challenge to the contractor to complete quickly, but may cost the owner more as the staffing levels are minimized by the contractor in order to maximize the return on the fee.

A fee structure that fixes the agreed profit and includes the contractor's overhead and indirect costs as part of the project costs allows the contractor to properly staff the project. Because the contractor is paid only the actual costs of the personnel charged to the project, it is more likely to use enough to do the job well because it is effectively not penalized for using the Owner's funds to staff the project adequately and, at the same time, is not rewarded for overstaffing it – as would be the case if it were paid more than its costs for the use of these personnel. This argument is flawed if the contractor can use its personnel elsewhere and be paid more. It is likely to assign no one or to assign people who are less than the best.

Cost plus fixed profit offers the contractor the opportunity to staff the project properly without being apparently penalized for doing so. This may or may not be of ultimate benefit to the buyer, depending in part on market conditions.

Just a bit earlier on in this section, I mentioned that there was some uncertainty about what constitutes direct costs versus indirect costs or overhead. This is based in part on a study undertaken in the mid-1990s on mark-ups for change orders. This study is discussed in more detail in Chapter 9. In summary, we found that there is no clarity as to what is intended by a number of headings that different types of items are supposedly covered by. This clearly creates an opportunity for games. We will look at some of the games people play at various stages in the life cycle of a contract later, in Chapter 9. That said, we do need to recognize that the opportunity exists to assign what one party considers to be covered by the "Plus" part of the contractor's compensation in a Cost Plus Contract to the "Cost" part. We also need to recognize that this can cause mistrust and – often as a result of this – grief.

 It is difficult to know what is really a 'cost' and what is not. There are many opportunities for creative accounting in cost plus fee arrangements.

The Wall Street Motivational Theory

This theory is simple. There are only two things that motivate people: fear and greed.

There have been many attempts, in various forms, to manage the behaviors of the contractor or vendor through incentive schemes. Most of these have their roots in the Wall Street Motivational Theory.

The incentives sometimes expose the contractor's profit and even its overhead to being lost if specific objectives are not met. In exchange, an opportunity to make a windfall profit is created if these targets are exceeded. The simplest versions tie additional earning potential to finishing earlier or to the project costing less than the target budget. This translates into increased revenue or return on investment for the buyer, who can then justify the additional payment. The simple versions have a big advantage in that they are easily understood. They have a disadvantage too, as they can become the focus of management activity, potentially to the exclusion of more important elements of the project. Or they can get in the way of achieving other objectives. One classic challenge is maintaining the right balance between capital and operating cost, as the savings that are often needed to earn the additional fee for the contractor will increase the cost of operation or maintenance of the finished product from the contract. From these types of internal conflicts have grown a range of more complex schemes. They tend to be harder to understand and therefore increase the potential for disputes later.

It is not unusual for cost plus contracts to be supplemented with an incentive scheme to manage the risks and to enhance performance by creating appropriate rewards for appropriate behaviors.

As with any endeavor, the stakeholders need to be happy with the result in the end. Keeping people happy is in large part directly linked to the ability of others to keep their expectations in line with what is going to happen. This seems to be particularly true of Cost Plus Contracts, in which expectations have a tendency to exceed the potential of either the contractor or the buyer to deliver.

The success of a Cost Plus Contract is closely linked to management of the expectations of all parties.

The importance of managing expectations is described in detail in "Don't Park Your Brain Outside" (Hartman, 2000).

Often, the choice of payment terms is (or at least should be) dictated by the availability of information that allows the contractor to accurately price the work to be done at the outset. A Stipulated Price Contract should be based on complete information that will allow the Contractor to price the Work accurately based on sufficiently accurate and reliable information regarding both existing conditions and expectations of what is to be delivered under the terms of the Agreement. If the amount of each type of work is less certain, then a Unit Rate Contract is a more appropriate choice. If the Work is ill defined or time pressure requires start of the Work before it can be fully defined, then a Cost Plus Contract works best.

The selection of a contract payment approach is often easiest to determine by looking at the amount and accuracy of the information available to the contractor.

Selection of the method of compensation for the contractor is one of the most fundamental steps in formulating a contracting strategy. This decision sets or determines many other factors and options. For example, a Stipulated Price Contract requires a contract selection process that allows the contractors who are being considered to assemble a price proposal based on a common and shared view of what is involved in successfully completing the contract under its various terms and conditions.

 Selection of the payment terms is the most fundamental – and therefore the first – step in formulating a contracting strategy.

The contracting strategy should also be aligned with the assessed and explicitly stated objectives of the project. Only after the overall contracting strategy is set can we look at how we bring the contractor onto the project team.

2.2 Tendering Options

Having looked at the three basic forms of payment terms, the next step in formulating a contracting strategy is to decide how the Contractor is going to be selected. In some cases, we already know whom we want as a contractor, in which case the process is simple: We negotiate. The contractor selection process serves several purposes. The first – obviously – is to pick the right contractor or supplier. The second – less obvious – is to make sure that an appropriate and auditable trail is left behind to show that the selection process was aboveboard. It is this second reason that is often the more important one, and strangely, it is this one that often gets in the way of picking the best contractor. We end up selecting the one that most closely meets our needs so it looks like we have made the "best" decision. This said, let us look at the ways in which we pick our contractors.

Contractor selection is not just about selecting contractors. It is also about looking like we have done the right thing.

When making a decision about whom you are going to work with in a contractual relationship, consider what you are buying. If it is a commodity (it really does not matter who provides it, it is the same thing), then select based on price. If there is any difference in the nature of what is being purchased (such as quality, speed of delivery, sustainable relationships, transfer of knowledge, competitive advantage, or people and their expertise or competence) then the decision should be made based on value. For example, buying a specialist's time is normally not a commodity decision. Well, maybe it is – but it should not be. A senior specialist may deliver a service or solve a problem in less time than a more junior or less experienced one may. The senior person costs $100 per hour and the junior one costs $50 per hour. If the senior person provides the same service in 1 hour that takes the junior one 3 hours, then the higher price per hour provides the best value.

In selecting the right vendor, value is more important than price if what is being purchased in not just a commodity.

We should never have hired Midway Engineers Inc.
to design this power station.

2.2.1 Bidding

This is the most traditional and reliable way of selecting a contractor. It does NOT necessarily give you the best contractor or the best deal, but it is the most readily accepted way of selecting a supplier. Usually, the criterion for selection is lowest price. When this is not the case, the process becomes much more complex.

Beware of the traditional bid process – it does not necessarily lead to the best choice of supplier for goods or services.

If the purpose of bidding for supply of goods or services is to obtain value, then the lowest price is not necessarily the best indicator that this will be achieved. Even when the buyer has specified what is required and has written a "watertight" contract for its delivery, the lowest price can lead to a higher final cost to the Owner. Consider the following hypothetical case:

We have selected Contractor A because it provided the lowest bid. Contractor B is known by us to be reliable, has done excellent work before and has delivered a quality product. Now consider the following events, adapted from a real case for contractor A. I have had to guess what could have happened had contractor B been selected, as we never know the outcome of something that did not occur! The example serves to illustrate how we may have picked the worst deal . . .

Event/Cost Item	Contractor A	Contractor B	Comments
1. Original Contract Price	$5,000,000	$5,650,000	Pick low bidder. Save $650,000.
2. Contract administration more expensive because of contractor's non-compliance to commercial terms	Marginal Cost: $35,000	$0	Need to spend more time in administration, processing of change orders, consultant reviews.

Event/Cost Item	Contractor A	Contractor B	Comments
3. Claims made by contractor for items that most others would have included in price	Marginal Cost: $345,000	 $0	Cost includes legal fees, additional consultant fees and buyer staff time.
4. Completion delayed	Marginal Cost: $2,500,000	 $0	Lost profit from sales losses. Excludes loss of goodwill.
5. Quality marginal. Lost production caused by additional shutdowns and maintenance costs – net present value over life cycle of project	Marginal Cost: $ 1,400,000	 $0	Cost of unanticipated maintenance, additional work by others to repair original, and lost production costs.
6. Legal fees and other internal costs associated with resolving disputes and minimizing other losses	Marginal Cost: $200,000	 $0	Cost of disruption during and after the contract resulting from the processes used by the contractor, threats of litigation and other risks.
Final costs	**$8,220,000**	**$5,650,000**	**Contractor B would have saved the buyer $2,570,000.**

Table 2-1. Comparing bids.

To avoid the situation that is illustrated (hypothetically) in the example above, we introduce other factors into the contractor selection process. Some of the things we can, and have, asked for include the following:

- Contractor references from clients
- History of claims and litigation
- List of key personnel and their resumes, references and possibly even interviews with key contractor staff.
- Project execution plan
- Risk mitigation plan
- List of key subcontractors and suppliers

When such additional items of information are requested, the contractors should be told how they will be assessed. In fact, in Canada, as a result of litigation in the early 1980s (referred to as the Ron Engineering Case), this is required if the buyer does not want to risk being sued for selecting *any* contractor other than the *lowest bidder*. From a purely practical point of view, it pays the buyer to be explicit in its expectations regarding what it is looking for in a contractor. From the perspective of being auditable, it also pays to be explicit in our contractor selection requirements and processes. Generally, good practice includes not just what is needed from the contractor to allow the buyer to fully evaluate the bids, but also enough information about the evaluation process to allow the contractor to understand how the decision will be made. There is an argument that if you give the contractor all of this information, it either ties the buyer's hands later or it allows the contractors to be too creative (dare I use the word "lie"?) in responding to the call for bids or proposals.

When tendering is not necessarily just to get the cheapest deal, then the rules for selection are best spelled out in detail to all of the bidders.

When asking potential vendors to bid on work, consider all of the factors that are important to the success of the buyer's venture. Often, value is more important than price. Despite this, there is a tendency to imply that price is the determining factor.

When value is more important than price, carefully list what constitutes this value, so the vendors can respond more effectively.

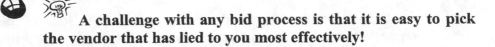 **A challenge with any bid process is that it is easy to pick the vendor that has lied to you most effectively!**

 Sometimes it is better to use a softer approach to vendor selection than the more traditional price-centric tender process.

There are as many ways to adapt the tender process to suit the specific needs of a project and a client as there are projects and clients. Unfortunately, much of the money spent by large corporations and government agencies tends to be paid to contractors selected using inappropriate selection processes. This is often because the process is seen as being too difficult to change to suit the specific needs that a given project, contract and circumstance dictates.

 The tender process <u>can</u> be adapted by the buyer to suit its specific needs.

It is generally accepted – and is really not in the least bit surprising – that we expect vendors to respond to our needs as buyers. This has been formalized to the point of stupidity in many organizations. The stupid part is that we now will not consider contractors (in some situations) when they do not fully comply with our explicit requirements as set out in the tender documents. This excludes creativity, innovation or better ideas. To address this issue, some buyers allow alternatives to the base bid (which must be compliant). Unfortunately, the buyer all too often will limit the options and opportunity for innovation by stating that it will only consider alternatives from the lowest bidder. If this is the case, the best ideas may not be put forward because vendors do not want the competition to have potential access to their ideas in the event that the owner shares those innovations or solutions with another contractor that happens to have submitted a lower bid for the base case.

Bidders generally need to be responsive to the bid requirements to be considered, although there are exceptions to this.

Owners may open invitations to any potential vendor for a contract, or they may limit the list of contractors or suppliers that may submit a bid. This is most easily done in the private sector, though it is not uncommon in the public sector. When the buyer wishes to restrict the number of offers to a manageable size or when the buyer does not want to entertain certain contractors' offers or if the requirements are such that some form of pre-qualification is needed, a two-stage bid process is common. The first stage (which may be entirely in-house within the buyer's organization or may involve prospective suppliers) is a pre-screening step. Only suppliers who pass this screening will then participate in submitting a proposal.

The tender process may be opened to pre-qualified bidders or to just about anybody. This is something the buyer must decide.

2.2.2 Request for Proposal

The difference between a Request for Proposal (RFP) and a formal tender process lies in the actual or implied request for additional information upon which a selection will be made. Let me explain. An RFP, as the name implies, is a request for some form of proposal to solve a problem. This implies that the solution(s) proposed by prospective suppliers form part of what the customer will assess in the selection process. This, in turn, implies that there is more being considered than just the price. It is good practice to specify clearly what is going to be assessed. This often includes the proposed personnel, proprietary technologies, technical solutions, project delivery plans and more. Increasingly, clients are looking for other "softer" information, such as referenceable projects of a similar nature and the track record of the contractor in claims, safety, labor relations and so on. Staff turnover, incentives and other traditionally internal-to-the-contractor factors may also be included as criteria for assessment.

Requests for proposals usually solicit a broad band of general information as well as project-specific solutions to help the buyer select a "best fit" supplier of goods or services.

We saw in Section 2.2 that buyers may ask for more information than just price. As this becomes more comprehensive, the type of contractor selection process shifts from tendering to requesting a proposal. This is a grey area, and is really not a formal distinction except for one factor. The important distinction is that we are sending a message to the contractor that we are looking for much more than just a price to deliver a predetermined solution. We are looking for the best solution. This is a significant shift, although the border between the two approaches is blurred in practice. In your RFP you should be very clear on what the proposal is being requested for. Often, such requests are muddied a bit by requiring the prospective vendor to provide information that is standard to the buyer's procurement process but totally irrelevant to the job in hand.

Requests For Proposals should focus on soliciting truly relevant information, and should reflect the other terms of the proposed contract.

One item of information that is often requested by potential buyers is evidence of the contractor's financial capacity to perform the work. This is an understandable request from a client's perspective as most buyers do not want the grief associated with a bankrupt vendor and all that it implies in terms of incurring delays, finding replacements, paying twice for sub-suppliers efforts and materials and so on. The challenge comes in the ability or even the willingness of vendors to provide the information requested by the buyer. Vendors, particularly those that are privately held corporations, may be reluctant to disclose the specific information requested by the buyer. If this is the case, the vendor will exclude itself from the contract. The buyer should consider carefully what it really needs, as some of the information requested may in fact not be necessary for what it is trying to achieve.

Financial competence to perform the work is often requested and sometimes is a barrier to contractors that are privately held.

Another fairly common request is for information regarding management processes of the contractor. This is of value only if the buyer really understands how the prospective vendors do their business. This information therefore has value if the persons requesting this information know how to properly assess it. A cynical observer may wonder if the prospective buyer requesting this information understands its own business processes well, let alone those of its prospective supplier. The time and effort required to evaluate others' management processes are considerable, which begs the question, "Is it really worth it?"

Owners seeking management information from their prospective contractors should know what they are asking for. They should be in a position to assess the management processes of their suppliers and they should be able to evaluate the accuracy of what they are being told.

Information requested from a contractor must serve a purpose. This purpose needs to be understood by the vendor if the vendor is going to respond in a way that will help it be successful. One of the natural responses for any contractor trying to win work is to tell the customer what it wants to hear! As one retired vice president of engineering of a large petrochemical company once commented to me at the end of a presentation on better contracting processes, "In the end, we tended to pick the contractor who has lied to us the best."

Contractors have a propensity to tell a prospective client what they think the client wants to hear . . . once the work is contracted, there is time to make adjustments to reflect reality!

The RFP process is probably best used to screen the vendors on a long list and to then interview and negotiate with the best or most interesting respondents. RFPs are more effective in helping identify whom you want to negotiate with, rather than being the ultimate step of actually selecting a vendor.

A well-prepared Request For Proposal, with good responses to it, often ends in negotiation with a short list of prospective suppliers or with the proponent with the best proposal.

It is not unusual to see an owner negotiate with two or more proponents.

2.2.3 Negotiation

Selection of a supplier and award of a contract may be based on consideration of a single potential source. Or it may be based on discussions with the best of a number of bidders or proponents who responded to an RFP.

Negotiation – in some form – usually precedes formation of a contract.

Even competitively tendered contracts usually go through a pre-award process in which specific items are "clarified".

Negotiation is not of any value if we do not know what we want! Negotiation is an integral part of any contracting process, even if we allege that we do not do it. As we work through the life of a project we end up negotiating changes, specific terms, interpretation of specific clauses, the meaning of specifications and more. Doing as much of this before signing the contract is useful as it takes advantage of the buyer's power at its peak (before the contract is signed) while allowing the contractor to correct any mistakes that are discovered in the process of negotiating and clarifying.

An important part of effective negotiation is to know what you really want.

 Another key to successful negotiation is to understand what the other side needs and how you can best deliver on that need.

As with everything else, contract negotiations require skill and an understanding of the "rules of the game". Although the idea of people playing games in contracting is not a politically correct one, it happens anyhow. The games start with negotiation at the outset of the contracting process. To illustrate this, consider one of the most common games: fighting for something you do not really want. This creates the impression that you give up a huge point – usually in exchange for an apparently smaller one. If the "huge" one is relatively – or even totally – unimportant to you, and the "small" one is critical, you win while the other party thinks it has won too. Played right, good negotiators will make a point of racking up a credit against a future favor in exchange, so they win twice!

 Contract negotiations tend to be fraught with games that people play. Understand these games and you can win while making the other side feel they have won too.

 A successful negotiation is one in which all players feel they have a good and (ideally) fair deal.

As negotiations progress on contracts, it becomes clear that more than just price and schedule are up for discussion.

Negotiation of a contract may affect scope, as well as terms and conditions of the final contract.

Negotiation at the outset of a contract is important not just for the purpose of obtaining the best deal, but also for the purpose of establishing the relationship between the contracting parties. This is a complex issue, as it is dependent on so many factors.

 Negotiation is not just part of the process of forming a contract; it is a critical part of establishing the relationship between the contracting parties. This relationship will influence the rest of the project completed under this contract.

2.3 Scope of Work

The scope of the work to be undertaken under the terms of a contract can vary significantly. The simplest deal from a buyer's perspective is to go to a single source for all of the requirements of a project. In the ultimate form, this can include everything from financing to implementation and even operation of the completed facility, service or product. As the number of separate sources of supply increase, so does the complexity associated with managing and coordinating the different contracts. The important point here is that the liabilities for failure of this coordination and management effort initially lie with the buyer.

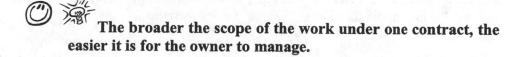 **The broader the scope of the work under one contract, the easier it is for the owner to manage.**

The greater the number of contractors required for a particular undertaking, the greater the risk and the effort required by the Owner to manage the project.

2.3.1 Supply Only

Contracting for the supply of an item or a commodity is perhaps the simplest form of contract. And the simplest form within this category is for procurement of a commodity that any supplier can provide quickly. In this case, the only consideration is price, assuming compliance with specifications and ability to deliver on time are not issues.

The easiest form of contract is to acquire goods. This is especially true when the goods in question are standard manufactured items or commodities.

In our day-to-day living we acquire products at the store and do not enter into a formal contracting process. There are many items that we acquire for business on the same basis. This includes cab rides, hotels and travel, printing services for small quantities, legal, accounting and other professional services, office consumables and small items from the hardware store.

 Supply-only contracts are often completed with no formal contract in place. In this case the parties normally rely on appropriate legislation or other established legal precedent (e.g., consumer protection acts or legal precedent) or accepted practices to govern their behaviors.

As the items we want to acquire become more specialized or specific, so the need to add to the process of contracting increases. Non-standard goods or services usually require at least a specification so that the vendor can understand exactly what we are looking for.

 Supply only of non-standard goods requires a specification that is clear so that the supplier can actually deliver what the buyer wants.

As the complexity of what we are trying to acquire increases, so does the potential for confusion as we introduce the opportunity for misunderstanding. Precision in specifications is remarkably elusive. Many books on contracting reflect the experiences of practitioners who have seen ambiguous or misleading or impossible-to-comply-with specifications. Classics include "to the satisfaction of the engineer", "as required by the architect" and my favourite, "as needed" – which leaves it open to anyone to declare a need!

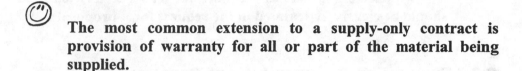

Ambiguity is a common cause of disagreement in contracts. Unclear specifications are one example where lack of clarity is common.

One of the most common extensions to any supply-only contract is a requirement that the product or materials being supplied will last. This is usually in the form of a warranty from the supplier or the manufacturer – not necessarily the same thing.

The most common extension to a supply-only contract is provision of warranty for all or part of the material being supplied.

Supply-only contracts are typically the easiest to manage and administer, because they involve the exchange of only tangible and definable deliverables (money in exchange for a product).

2.3.2 Services Only

The next most complex category of contract scope is for supply of services (such as design, manufacture, shipping, or construction). Services tend to be harder to specify, as they often have a partially intangible component. Intangible components include the following (an incomplete list):

- Efficiency of the contractor
- Skill of the specialist
- Quality of service
- Responsiveness to customer needs
- Ability to understand the real needs of the client
- Effectiveness of communication
- Availability of key personnel
- Effectiveness of contractor processes
- Currency of technology and knowledge
- Innovation and creativity.

The items on this list have two common features: They are hard to measure and are therefore difficult to define, deliver or manage. And they are all dependent on people.

Services involve provision of less tangible products that are harder to measure and are therefore harder to assess in terms of value.

In selecting a service vendor, the criteria for selection should be clearly articulated in the request for a proposal.

2.3.3 Engineering, Procurement, and Construction (EPC)

> **Note**
>
> The headings in Chapter 2 refer to terms that are commonly used in the construction industry. This is because that industry typically developed the concepts and approaches. The approaches are used in other industries too, which may use different terminology.

Combining two or more services is quite common. Combining engineering with procurement of the engineered components and then their assembly is a natural combination that integrates elements that are best handled by a single vendor. The engineering, procurement, and construction (EPC) contractor is set up to provide this integrated service. It makes a lot of sense if the effort in each of the three areas is likely to be intense and need close integration to be carried out efficiently and quickly.

Most commonly used in the petrochemical industry, Engineering, Procurement, and Construction contracting offers a packaged approach to the design and delivery of technologically complex facilities.

Although this concept is readily recognized in the petrochemical industry, it is also used in other industries to provide integrated engineered solutions to the supply of process facilities to the automotive industry, pharmaceutical process and manufacturing solutions, as well as software or systems implementations.

In other industries, the equivalent of EPC contracting exists. Generically, this approach is a bundled service providing technology integration and delivery services through one prime contractor.

Although not only delivered through a Cost Plus contract, this approach is often well suited to situations in which the solution cannot be defined until well after the contractor has been selected. For this type of service, it makes sense to select a contractor that has the expertise and the processes in place to deliver what the client wants. It makes sense to select on a criterion other than price, but the selection process is often reduced to price only through a pre-qualification process that reduces the list of bidders to those who have the potential or apparent capacity to do the work.

Because of the nature of this approach, it is most commonly provided based on a cost-plus contract for payment.

Buyers need to trust the contractor, its experts and its capability as manifested through its business processes, procedures as well as the business and technical approach it uses. This is because much of the work done and many key detailed decisions made on behalf of the buyer will be controlled by the contractor. It used to be that many owners retained an almost complete parallel staff to the contractor in-house. Those inefficient days are gone, so reliance on the professionalism and integrity of the EPC contractor has grown considerably.

The Fastest Drawing in the West

The project manager finally had enough. He rolled up the drawings and took them with him to his client's design review meeting. The client unrolled the drawings. Apart from the title block, all were blank. The client turned to the engineering, procurement, and construction (EPC) contractor's project manager and asked him, none too gently, why he was wasting his client's time with a set of blank drawings.

The EPC contractor countered that the drawings were not blank, but that they had complete title blocks. He continued by pointing out that, so far on the project, all the drawings that his engineers had produced were completely revised by the client's parallel team of engineers. He was saving the client precious engineering time and fees by cutting out the middleman.

He made his point. From then on, the client relied on the contractor to complete the design. The client reduced its staff considerably and never looked back on the decision not to duplicate the work of its EPC contractor.

Corporate and individual reluctance to trust third parties is rampant and needs to be overcome by the buyer organization and the individuals in it charged with the successful delivery of the project. Sometimes this is virtually impossible, so the owner resorts to substitutes to this type of trust (see Overkill).

Overkill

This large multinational petroleum company had found the ultimate solution to working with an engineering, procurement, and construction (EPC) contractor that was to design and build a new refinery for them. Their experience had been that nameplate production was rarely achieved at plant start-up, so they required a significant amount of effort to "tune up" the facility after the EPC contractors had left. This was expensive and time consuming and affected throughput and thus revenues from the facility. Their solution was to tie performance bonuses and penalties to start-up performance of the new refinery. At first glance, it worked. Start-up went well, and the $240 million plant met all production requirements from the outset.

The mistake became apparent some ten years later when the capacity of the plant was doubled for less than $80 million. The reason: gross overdesign of the original refinery. The reason for this overdesign? It is better for an EPC contractor to spend the customer's money than risk having to pay a penalty. An estimated $80 million worth of facility was invested in with no real return on this investment until the refinery was de-bottlenecked a decade later. Clearly, the solution created a bigger problem than the one it solved.

From a buyer perspective, a significant amount of trust is needed for this type of contract arrangement. Substitutes for trust are sometimes used, but these can cost the buyer more than what they may save.

EPC contractors need to be able to work within the parameters set by the buyer of their services. Simply put, the contract type and terms need to reflect the reality of the project. Even a Cost Plus Contract can be unattractive and too risky if the project is undefined and the opportunity to earn sufficient fees is dependent on an unknown scope of work. A fixed fee, for example, makes little sense to an EPC contractor if there are too many unknowns to assess this fee, let alone provide a reasonable budget for the reimbursable portion of the contract.

From a vendor perspective, EPC contracting carries a significant risk if the fee structure is based on cost plus a fixed fee and the scope of the project is poorly defined at the time of contracting.

The nature of the project evolves from poorly defined at the outset to well defined as it approaches completion. As this happens, the appropriate type of contract effectively changes. To handle this transition and all that it implies, some fairly complex contracting models may be used. One such concept is the use of an evolving contract. This is a contract that typically starts based on a reimbursable format and is converted at some appropriate point to one that is incentive based when reasonable targets can be defined. It may even evolve later into a Stipulated Price Contract.

Fairly complex fee structures may be required to obtain an equitable contract for supply of services to design, procure and build a facility, system or other project.

Incentives for performance are used in some instances. The design of the incentive scheme needs to be carefully considered and aligned with the needs of the project.

This type of contracting approach puts most of the control for execution of the project in the hands of the contractor.

2.3.4 Engineering, Procurement, and Construction Management (EPCM)

Engineering, procurement, and construction management (EPCM) contracts are a variation on the EPC format. The difference lies in how the construction portion of the work is contracted. Construction is the costliest part of the work. There are several reasons why the buyer may wish either to defer commitment to the construction phase or may wish to retain greater control over it. The EPCM variant allows the buyer a bit more flexibility and control.

EPCM contracts are a modification of the EPC arrangement under which the buyer takes on direct payment to many of the specialist contractors involved in the construction phase of the project.

Typically the coordination of construction with design and procurement under an EPCM contract remains with the EPCM contractor. The individual construction work packages are developed and managed by the supplier of the design and procurement services, who now acts as the Construction Manager (CM). The buyer typically makes the final selection of the individual construction contractors and retains the contractual relationship with them. This may cause a degree of exposure too, if the overall contracting strategy is not well aligned (see So You Think You Got a Deal).

The construction management portion of this type of contract makes the supplier of the design and procurement services responsible for the coordination of the construction or implementation phase.

Under this type of arrangement, the buyer typically contracts directly with the contractor(s) for the construction of the work, but these contractor(s) are managed and coordinated by the EPCM Contractor.

From a buyer's perspective, this approach has the advantage that the buyer retains the contractual relationship with its suppliers while passing the responsibility for managing and co-ordinating the work to its EPCM Contractor.

From the EPCM Contractor's perspective, the advantage of this approach is that it can effectively retain the control needed for more effective delivery of the project while not having to tie up much of its operating capital because the buyer pays the specialist contractors directly.

So You Think You Got a Deal

In the days when I ran a construction company, we bid on a work package for construction of some foundations for a large factory. The package was to be managed by an engineering, procurement, and construction management (EPCM) contractor, but the contract would be directly with the owner of the factory that was being built. Design was underway, and it appeared that the EPCM contractor's contract included onerous professional liability terms.

The bid documents were standard for the industry except for one clause added by the EPCM contractor that allowed them to delay the work at any point in time and for any amount of time and for any reason. Our price was to include for such delays. And making up the time against the target completion date was also at our own cost. We saw this as a huge risk. Accordingly, we bid $5.2 million but offered to do the work for $4 million if they removed this one clause from the contract. The EPCM contractor asked us to meet with them. They told us that our bid was the lowest at $4 million, but that we would have to accept the clause covering any delay. We refused. They said that, in this case, our low bid was noncompliant and we were out of the competition.

(Cont.)

> ## So You Think You Got a Deal (cont.)
>
> I heard later from the successful contractor that they had won the job – with the clause in – for $4.8 million. In other words, the inclusion of one clause to protect the EPCM contractor from a perceived professional liability risk had cost the owner $800,000. I saw no real delays as I drove past the site over the following months, so I assume that this cost turned into a significant windfall profit for the successful construction contractor.

There is a link between control and risk management that all buyers need to be aware of.

2.3.5 Design/Build

Design/build contracting has much in common with EPC contracting. Its roots are in the commercial construction arena, where many similar buildings are produced and the processes are well understood. Often, design/build is associated with a real estate deal as well. Many of the warehouses and light manufacturing facilities that spring up in industrial subdivisions around a town in North America are built using this approach. Typically, these buildings are large open areas surrounded by a building shell with some office space in the front. Different treatments of the elevations make remarkably similar boxes look like different buildings.

The design/build contract model has its origins in commercial construction.

The primary way in which design/build contracts vary lies in their scope. For example, a real estate developer using this approach will often offer for lease a building designed to suit the tenant. The deal then includes land, design and construction, financing and a lease – usually long term – to the end customer. The developer may structure these projects for the purposes of holding the property or of transferring it to a long-term investor, such as a life insurance company or a pension fund looking for a longer-term investment with a degree of security. Parallel deals exist in the use of technology. Systems may be leased. The assembly of the hardware, software and infrastructure (computer room, intranet, local area network, etc.) may all be contracted out to a specialist firm.

There are many models of design/build that are used. These models tend to vary the scope of the work to be done or the services to be provided.

Especially in situations in which the customer is a person or organization that is not normally in the business of assembling the components required to obtain a facility, a deal is struck based on a performance specification rather than a detailed technical one. What the buyer is looking for – and hopefully the vendor is offering – is the ability to provide space and capability for manufacturing something. Alternatively, this could be a facility or capacity, such as computing capacity and related support services that the buyer needs to do its business.

It is quite common for design/build projects to be based on a performance specification rather than a detailed one.

One of the primary attractions of the design/build formula is that it is one-stop and (theoretically) trouble-free shopping, in which the buyer knows up-front what they are going to get and what it will cost. For this to be a reality, the fixed price needs to be based on a fixed target. This target is normally defined in terms of a specific scope and quality to be delivered in a certain timeframe.

Many design/build contracts are based on a fixed price for a · fixed and defined scope and quality of product.

From a buyer's point of view, this type of contract model serves to provide a single-point of accountability. The risk lies in the adequacy of the specifications.

An important feature of the design/build process is that it allows designers to play to the strengths of the builders. Perhaps more correctly, it allows the builders to influence the design so that they can take full advantage of their strengths, including their preferred suppliers and specialist contractors.

From a vendor's perspective, the advantage of this approach lies in being able to detail the end product to suit the vendor's capabilities as both a buyer of external goods and services and as an integrator and assembler of the end product.

2.3.6 Build, Own and Transfer (BOT) and Build, Own, Operate and Transfer (BOOT) Contracts

We have just seen that the design/build contract form allows for scope adjustments that include financing. One step beyond this is for the contract to be with a customer that wants a facility but that cannot afford to own it – at least for a while. The build, own and transfer (BOT) and build, own, operate and transfer (BOOT) contract forms were developed to address this market need. Expansion, or even survival, of an organization may depend on providing facilities, goods or services that the provider cannot afford to pay for directly. In some situations, these organizations cannot compete in providing a product or service to their own customers, as well as someone else. Alternatively they may simply be looking for off balance sheet financing. The degree of vertical integration that is needed then expands to include off balance sheet financing and operation. If the scope includes operation for a specified time, a BOOT contract is used. If no operation is included, then a BOT contract is used. The package for a BOT contract is usually based on a guaranteed or underwritten revenue stream from the user for a period of time after which the user may purchase the facility for a specified (often nominal) amount. The transfer of the facility is usually linked to expected conditions surrounding that transfer, so that the ultimate beneficiary is left with a useful product. If the ultimate beneficiary operates the facility during the operating period of the contract, then there are usually standards set to which the facility must be maintained or to which the facility is deemed to have been returned upon transfer or breach of contract.

 BOT and BOOT contracts provide a comprehensive packaged service and product to the client.

Why should anyone go for such a complicated approach to obtaining a facility? The primary reason is that the alternative is to tie up operating or other capital, if available, or to obtain a loan or to float shares to cover the investment. Tying up capital may preclude other investment opportunities that are harder to finance or that will otherwise limit the growth and operation of the business or organization. Issue of new equity may take too long, may dilute the holding of existing shareholders or simply may not be an option. And borrowing the money – if the option is available – will appear as a liability on the balance sheet. These options affect the financial statements of an organization in a significant way. Obtaining what is, in effect, the use of a new facility without funding in the traditional way creates an opportunity to gain the effect of a cash infusion to the business without affecting the balance sheet.

Ownership of the facility by a third party is what enables financing of this type.

 The buyers of many BOT contracts provide off-balance sheet financing for capital improvements.

BOOT Contracts have an extra "O" for "operation". The deal is essentially the same as a BOT deal, except the owner of the facility operates it on behalf of the ultimate beneficiary for a number of years.

 In addition to what a BOT contract provides, a BOOT contract includes an operations phase, which moves part of the operating cost of a business to the contractor in exchange for a revenue stream to the contractor over an extended time period.

A Golden Opportunity illustrates how one deal was put together.

A Golden Opportunity

A West African company that owned a mining concession had found gold. They wanted the mine developed and a refining facility built. They had the site of the gold pinpointed, a local population that knew little or nothing about gold recovery, and an ambition to develop an opportunity.

A financing syndicate was assembled, and an operating company was established that would operate the mine and plant for a 5-year period. The development of the mine and the plant was financed based on selling gold futures linked to a guaranteed production from the plant. The reserves of gold in the mine were underwritten by an insurance consortium based, in turn, on an independent assay.

The plant was built based on the money raised from the sale of gold futures, and the facility was operated under a build, own, operate, and transfer agreement for 5 years. During that time, any surplus production revenues (above the requirements to meet the repayment of the loan based on the gold futures) were shared with the end beneficiary – the owner of the mine and mineral rights. Also during that period, the locally hired operators and staff were trained to operate the mine and plant. At the end of the 5-year period, the facility was to be handed over to the owners, who would operate it from there.

2.3.7 Turnkey

Another variation on the general theme of bundling and vertically integrating services is the Turnkey Contract. This type of contract provides everything that a buyer needs to obtain a new facility up to the point where the key is handed over to the buyer. The buyer only has to pay for the product and turn the key to get it operating.

 Turnkey contracts provide a bundled package usually for a lump sum stipulated price.

There are many variants of this type of agreement. To what extent will the ultimate owner be involved in the project financing? Does the buyer want to have its own staff trained in the operation of the facility? Does the buyer expect the contractor to commission the facility? What is the warranty commitment after transfer? Even with these relatively minor variations on the basic concept, there are some clear parallels and common areas with the BOT and BOOT contract models.

 There are many similarities in Turnkey, BOT, BOOT and Design/build Contracts.

2.3.8 Other Flavors

There are as many variations in the ways people contract as the human imagination allows. This helps us understand that there is no practical limit to the way we can bundle goods and services, how we arrange to finance and pay for such things and what the scope of a contract will cover. Sometimes old labels are recycled for such modifications to established contracting models, and sometimes new terms are coined.

 There are more ways to contract than we will ever use, consider and possibly know about in our careers.

The "right" model of contract for a particular project is one that best addresses the needs of the contracting parties. We can look at these needs under three headings.

First, we need to consider what the buyer wishes to acquire. This comes under the heading of "technical needs". Then we consider what the viable commercial terms are that will allow the buyer to acquire this item and still be competitive. These are the "business needs". The vendor needs to be able to align with this in order to respond to the client's needs. Finally, we need to consider the concerns, issues, policies and other factors that affect or govern the behaviors of the people involved. Considering "social needs" covers this.

 The only limitation to variety is the imagination of the people trying to find the right balance among business, technical and social issues.

2.4 Vertical Integration

Increasingly, as businesses redefine themselves to provide increasing value to their customers by offering goods or services that integrate others' products, so we see a trend toward vertical integration. This is reinforced by the need to continually redefine our core business. Most competitive businesses define core competencies quite narrowly in a particular area of endeavor. This means that there is an increasing need to outsource what used to be in-house expertise. This need arises for a number of additional reasons, not least of which is the burgeoning amount of technology that is available. This growth in the number of technological artefacts (it doubles every 3 years by some estimates) means that we can no longer keep current with all of the relevant technologies in our business. Once again, this creates pressure on outsourcing.

There is an increased need to outsource what used to be considered core competency in many technology-based businesses.

The role of vertical integration becomes a natural niche for suppliers to fill and buyers to seek as the amount of contracting increases. Vertical integration is the process of linking elements in a supply chain in order to reduce the number of contact points. In other words, I used to deal with the supplier of six components and an assembler of those components. Now I deal with the assembler only and that company contracts with the suppliers to meet my needs. We see this happen every day in the construction industry in which a general contractor retains a number of specialist trade contractors to meet its obligations under a prime contract with a buyer. Similar relationships are common on the design side of this industry where owners contract with an architectural firm for the design of an office tower (for example). The architectural firm will subcontract the work for all of the other disciplines that it does not have available in-house. This could include geotechnical expertise, a number of engineering disciplines from structural and mechanical engineering to electrical and building skin design specialists. In addition, interior design, landscaping and other consultants may be retained.

 Vertical integration is all about packaging components that are traditionally separated by the client/supplier relationship.

It is only one more step to start to extend this concept to integrate financing, design, construction, operation of the building, and more, when none of this is part of the core business of the buyer organization. Additional incentives beyond just operations considerations also exist (e.g., creative financing). In the example of the office building, the customer may single-source the deal by signing a long-term lease that includes maintenance and other services. This deal may be signed with one vendor that in turn arranges for the financing, land acquisition, design, permits, construction and building maintenance through its own resources or through the use of a series of separate contracts.

 There are many ways of integrating contracts vertically.

2.4.1 Financing
One of the growing areas for vertical integration is the provision of financing that helps the client organization either through tax advantages or through better financial ratios. Leases are replacing capital investments. In some cases, when projects are funded primarily on cash flow, this source of financing falls short of demand from time to time.

For example, rapidly expanding populations create a demand for infrastructure that the tax base of the community involved cannot afford. An obvious source of financing is from the private sector. Within the private sector, investors come in many shapes and sizes. Apart from individual entrepreneurs and traditional sources such as banks, other groups include pension funds, insurance companies and venture capital firms. Other possible sources of funds include larger companies trying to diversify. For example, many companies are merging, acquiring and shifting direction. As they do so, opportunities for appropriate investment in other projects are a reality. Contractors, too, may be a source of financing as they increasingly consider taking all or part of their fees in equity in the project they are constructing.

A Convenient Bundle

This project was massive. One important but relatively small component of it – valued at a capital investment of about $300 million – was a pipeline to transport product from the production end of the facility in a remote location to the marketplace at the collection point of another pipeline system.

The owner of the overall project – in reality, an alliance of several large companies – decided to go out for proposals from pipeline operating companies to build and operate the pipeline for a number of years. The opportunity lay in the relative expectations of the different companies' shareholders regarding return on investment. The pipeline company wanted secure income, whereas the consortium wanted to maximize the return on investment and was willing to take some risk in doing so.

The successful pipeline company signed a deal to build and operate the required pipeline. It would meet the production dates and would guarantee the performance of the pipeline over the specified duration of the deal. In return, the consortium guaranteed a minimum throughput of product and a rate for shipping. The pipeline is financed, designed, built, operated and maintained by a pipeline company but it is an integral part of the larger project.

The pipeline has increased its revenues over the life of the project with a guaranteed cash flow at an acceptable rate of return. The consortium has financed a substantial part of its project through off balance sheet funding and has got rid of what it perceives to be the headache of operating the pipeline itself. Everyone wins.

Involvement of businesses that can add value, not just money, is of growing interest to potential partners with investment opportunities. Power co-generation projects are a prime example of this type of thinking. Situations arise where there is an abundant – or at least adequate and fairly reliable – source of energy and there is a facility producing heat and effectively throwing it away. It makes sense to harness that heat and turn it into electrical energy, then generate a revenue stream from this commodity. Often, the expertise to do all the additional work is not available in the organization that owns the base facility, so a co-generation partner is brought in. That partner may invest in an equity position in the deal in exchange for the opportunity to sell the electrical energy. The structure and details of the deal can vary considerably from one project to the next.

There is an increasing variety of sources of funding for projects.

2.4.2 Subcontracting

Subcontracting parts of the work taken on by a vendor under a contract is common practice and it is probably the best understood form of vertical integration. Most subcontracts will incorporate some or all of the key terms and conditions of the primary contract directly or by reference or implication.

Subcontracting is the most common and probably the oldest form of vertical integration of contracts.

One of the most important elements of subcontracting effectively is the coordination of the subcontractors so that none are held up or compromised in some other way in their delivery by any other contractor. Such compromise could lead to damages, potential disputes and other problems that ultimately affect the efficiency of the prime contractor. Essentially, the prime buyer has passed on this coordination function to the contractor by defining the scope of the contract to include all the work of the subcontractors.

Coordination between subcontractors is key to the success of the prime contractor.

2.4.3 Equipment and Material Supply

Each industry and industry sector has its own traditions regarding who procures equipment and materials. In the construction industry, for example, there are two quite different and fairly universal traditions. In the heavy industrial sector (refineries, heavy manufacturing, petrochemical plants for example) the procurement is done largely by the owner or by its EPC or EPCM contractor. The prime construction contractor, if there is one, still obtains the purchased items from the client or its representative. In industrial, commercial and institutional (ICI) construction, the construction contractor does most of the procurement of equipment and materials. Much of the reason for this difference lies in the lead time required for manufacture and delivery of specialist items that are purchased. Thus, the ICI sector tends to be differently vertically integrated than the sectors that rely on EPC or EPCM contracts.

 Different industries and industry sectors have different traditions regarding procurement of equipment and materials.

Users of EPC or EPCM contracts usually do so because a significant part of the project (and therefore the cost) is in bespoke equipment, designed specifically for a specialized task. This can include large machinery (e.g., ship loaders, paper-making machines, process vessels) and specially built or adapted devices (such as controllers, valves, conveyors). These take much longer to deliver than off-the-shelf components. The latter are more commonly – though not exclusively – used in ICI projects.

A primary driver for the difference in approach is the time required for manufacture or fabrication of components and their delivery to the final destination.

2.4.4 Operating

We have looked at the potential role of the vendor in the operation of a facility that it delivers as part of the process of vertically integrating the components of a project into one deal. Adding the operating component has an interesting effect, as it creates a real case for life-cycle cost optimization. Adding this part to a contract can help ensure that the optimal balance between capital and operating cost is obtained. Adding operations to the deal puts the contractor in a position in which it has a vested interest in the effectiveness of the design and the reliability of the finished product.

 Inclusion of operations as part of the package adds a commitment by the contractor to the reliability of the product.

2.5 Horizontal Integration

Whereas vertical integration considers the combination of different parts of a supply chain to gain advantages for the buyer, horizontal integration combines parts of a project that are normally separated in time or by skill set.

Most of the more traditional approaches to contracting are built on the presumption that we complete one step before going on to the next. Today's time pressures often preclude this type of approach. We need to start on the next phase before we have completed the current one.

Horizontal integration is an inevitable fall-out of the growing pressure to do things faster.

2.5.1 The Effects of Mixing Some or All of the Steps in Delivery of a Project

There is a reason why we do things in a particular order. We like to understand what we are to design before we design it. We can procure materials and equipment only if we can specify them. We can build something only if we know what is to be built. Traditionally, the functions of supplying materials, manufacturing and delivering equipment and assembling the parts fall into different companies' purviews. Similarly, in several industries we have separated – through evolution – the design function from those that follow. This is abundantly clear in construction. Originally, we had the master builder. This was when most buildings were essentially stone and wood fabrications (castles, churches and cloisters for example). Then we became more sophisticated and introduced all sorts of innovations – like indoor plumbing and electricity. So we specialized and segregated the functions. Now we are re-assembling them, being driven by the need to make things happen faster. The master builder is back in a much more sophisticated form than before.

As we increase the concurrency of previously segregated phases of a project, we introduce new challenges in management. Different disciplines are now working together. Greater coordination is required. The lines between functions become blurred. Professional liability issues become more important – or at least more sensitive. The obvious advantage gained in exchange for this added risk is that the project gets completed more quickly, allowing the owner to use it and, where this is relevant, to earn revenue from it to start paying for the investment.

I see this is a low-risk contract.

As soon as we move away from conventional contracting strategies, we take on additional risk. This additional risk needs to be offset by a benefit or increased value in some form.

Owners, as the buyers of design services, equipment, materials and construction services, have significant control over the entire process. In the construction industry, the designers (the Architect or the Engineer) have the next highest level of control. Often, they represent the Owner's interest and may even act as the Owner's Agent. As we overlap design with procurement and construction, so the control shifts. To be managed effectively, fast-tracked projects are driven by the construction schedule, not by the logical sequence of design. We do not normally design in the same sequence as we build. More on this in a moment. The shift in focus gives the Contractor doing the construction more influence and will change the balance of control so that the Owner and its Consultants have less effective control than before. Equally, if several specialist design consulting roles are now required to work more in parallel than in sequence, the need for coordination grows and this too muddies the waters regarding who does or controls what.

Loss of some or all control over the contracting process and the delivery of the project is a natural outcome of horizontal integration.

One of the most obvious solutions to the need to coordinate more between companies is for the Owner to select one of the companies as a Contractor and expect that organization to take the risks, rewards, obligations and headaches of this coordination on its behalf. This takes the significant task of coordinating the needs of one contractor that are supplied by another contractor away from the buyer. This coordination task needs considerable knowledge of the entire process and of how the different contractors' work affects others in the process.

Combining different elements of work that are normally separated by time usually means that the contract will require traditionally separate services to be integrated by the vendor.

When a vendor offers to provide an integrated service that spans several disciplines and services that have traditionally been provided by separate vendors, the management of the overall project requires a significant level of understanding of the specific tasks and processes required by each of the independent suppliers to complete their work. The key lies in knowing where the interfaces are and ensuring that the right information or product is passed from one contractor or vendor to another. This is no mean trick!

 Horizontal integration typically requires an additional level of sophistication from the vendor.

2.5.2 Fast-Tracking

The obvious place to start with fast-tracking of construction projects is to overlap the design and construction phases. Similar points in other industries lie in product development, where manufacturing processes are being developed while the product is still being prototyped or tested. In software development, we see detailed design happening concurrently with coding and testing. Each of these approaches requires packaging of the work so that enough of the preceding phase can be completed to allow the next phase to begin.

 One of the most common mixes is to fast-track design and construction.

Fast-tracking any project requires second-guessing parts of the work because the sequences are often modified. Consider design of a process plant, for example. We start with process design, followed by development of the engineering for the process – vessels, pumps, tanks and so on. This is followed by piping, instrumentation, power supply and finally the structure that supports all of this and houses it. Only then do we have the loading that allow us to correctly design the foundations. Yet when we fast-track, we need the foundation design first, as that is what we start building. So we normally guess at this. Then we guess at the structure and so on, as necessary, to permit construction to start. These guesses are not random, but are based on knowledge, experience and usually carefully documented assumptions. The guesses, however, sometimes turn out wrong, so the design needs to be reworked. If the timing is unfortunate, we need to rework parts of the construction too. Long delivery items also require some second-guessing by the designers, who cannot be expected to predict the thinking and designs of all of the others involved in the process. Management of the sequencing, packaging and modularization of the final project is both an art and a science that requires detailed and rigorous knowledge of both the design and the construction processes involved, as well as a wealth of experience. This experience accumulates only towards the end of a professional's career and then they retire!

 The biggest challenge in fast-tracking is management of the amount of rework or wastage that will result from wrong guesses.

Unless you have spent some time working in a fast-track environment, the "backwards" process for design to match the needs of construction can leave you wondering whether you are doing things intelligently. Certainly, the process is non-intuitive, based on the training we have as engineers, architects or other related professionals. Over-sizing the parts and understanding the economics of either speed and design flexibility versus additional materials or use of alternative construction methods or odd sequencing of work to facilitate the fast-tracking process are not topics covered in the average university degree program for that profession!

Doing work in a non-intuitive sequence often leaves people a bit confused.

One of the results of the inevitable modifications to the normal sequence of working in fast-tracked projects is that we over-design early construction elements. This is to allow a degree of safety to cover unknown loads on foundations or location of loads on structures. We may over-design bearing pads to allow greater tolerances for other designers so they can adapt to fine-tuning of process elements. At first blush, this can look like wasted money to the buyer. Explaining the rationale (risks of rework, need for flexibility to get the most important parts right and the related time-cost trade-offs) is an important part of the project manager's role.

 When fast-tracking a project, customer expectations need to be managed more carefully, especially if that customer is not used to this approach.

Back-to-Front Projects

A new building was to be constructed this summer. The ice bridges to the remote arctic location were going to break up in about one month. The structure would take about eight weeks to design. If the steel were fabricated after the design, the material would have to be air freighted up to the site if the building was to be erected this year. The solution was to ship by road "roughly what we need in the way of materials – with a safety margin" before doing the design. The design would then be completed based on available materials on the site.

On another project, at a remote location, the foundations were poured before the holes were dug for them. Here is why. The contractor found out that there was a shortage of carpenters, and the few who were available would require huge premium wages to go to this site. So the decision was made to build the foundation formwork at a city location about three hours away. The foreman pictured these large forms going up on a truck and saw that they were empty. What a waste of space, he thought, so he suggested adding the reinforcing steel for the footings and carrying them inside the prebuilt formwork.

(Cont.)

> ## Back-to-Front Projects (cont.)
>
> To get approval, the foreman spoke to the engineer, who now saw another problem and a better solution. The other problem was the cost of concrete at the remote location. And the forms – even with the rebar in them – were still pretty empty. So, they decided to add the concrete at the source in the city too. Now the precast footings would be shipped to site. The holes would be dug and backfilled to the level just below the foundation, with engineered lean concrete to substitute for undisturbed soil. Although the shipping cost was higher, the overall cost was significantly lower as a result of being a bit creative with the normal sequence of doing work.

2.5.3 Concurrent Engineering

There is an important difference between fast-tracking and concurrent engineering. While fast-tracking is all about getting things done faster by overlapping activities, concurrent engineering is all about applying the right expertise at the right time. It comes from the manufacturing sector. There were probably two drivers that led to this approach: faster time to market, and better supply chain integration. The end result, when it is applied well, is a better product, cheaper and faster. This is achieved by harnessing the expertise of specialists – typically from suppliers of components of the end product – during the design phase of the project.

The concept of concurrent engineering comes from manufacturing. It is all about getting the right expertise at the right time in the development of a design.

As the technical world continues to become more complex, so the need for specific focused expertise grows. We cannot know everything (even if this thought is bad for our technical egos) and, as a result, it often pays to get a real expert involved. Who better to help us specify the liner for a pump, or the coating for some steel, than the people who manufacture the product in question. They have, after all, seen their product used many times over and probably understand its application better than we can ever hope to as designers. The understandable concern is that the expert has a vested interest in selling his or her product. If you know your business, you will ask the right questions and should be able to determine the veracity of the expert's advice fairly easily. If you do not know the business and are concerned about the application or the product, you can always look for recourse through a warranty or contract term.

A big benefit of concurrent engineering is the ability to harness the expertise of specialist suppliers, who can help support more effective value engineering.

The need to test and validate the trustworthiness of the expert's advice is possibly just part of the issue. Trust plays a big part in fully harnessing the benefits of concurrent engineering. You may be in a competitive situation or your vendor may be competing for the contract. So sharing proprietary information – in either direction – will require a high degree of integrity by both parties. Further, the approach used in the contractor selection process will need to be thought out carefully in order to protect such proprietary or competitive information from leaking out. The best safeguard against this is the need for future business relationships. If such need to do business in the future exists, the relationship becomes more important and therefore greater care is likely to be taken. Put another way, we probably need to protect and nurture both the potential for future business and the track record of treating our business partner fairly and professionally.

The challenge with concurrent engineering as described here is that a high degree of trust is required by the vendors who are providing expertise and proprietary knowledge to deliver greater value for the customer.

2.6 Multicontract Projects

Many of the examples of contract types we have looked at in this chapter involve more than one contract for a given project. The contractual relationships between a buyer and its supply chain can be quite complex. The typical types of relationships that may be formed between a buyer and others to deliver a project are illustrated in Figure 2-1.

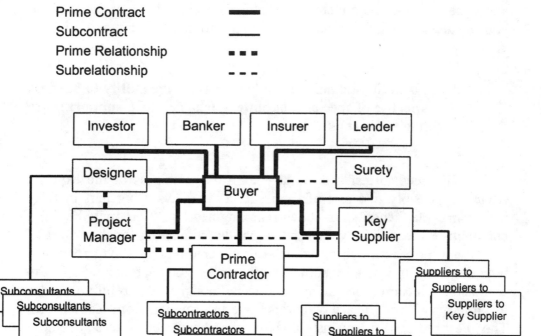

Figure 2-1. **Sample of simplified contractual relationships for a project.**

Figure 2-1 shows a simplified set of contractual relationships. Only the main contracts and contractual relationships are shown. The term "contractual relationship" in this context means that a contract specifies a role for one party that affects another party. For example, the Project Manager may act as Agent for the Owner in its relationship between the Designer and the Contractor. The Surety has a Contract with the Contractor (the policy) but will compensate the Owner in the event that the policy is called upon successfully.

There are many and often quite complex, contractual relation-ships on a typical project.

One of the big challenges that arise on projects in terms of efficiency of contracting is that these contracts are not normally coordinated. They have no alignment of purpose and they may even be in conflict with each other. We are so used to this situation that we assume it to be both normal and something that we should not particularly bother about. Now consider the following common situation in the construction industry.

A buyer enters into a contract with a company to build something. This Contractor retains part of the Work to do itself and subcontracts other parts. Specifically, this is a building that has an overhead door and a sprinkler system in it. An architect designed the building for the buyer. The Architect, as is common practice, retained separate specialist electrical and mechanical engineers to design the parts of the building that came under their jurisdiction. The overhead door is electrically operated. The sprinkler system is connected to a central security system that is designed to close the overhead door if it is activated. The electrical door opening mechanism is then supposed to be disabled and left so that it can be opened manually from either side. The electrical drawings point to the necessary wiring for the overhead door and state "Wiring by others". The mechanical drawings do the same. The building is built and the door interlock system with the sprinklers does not work. Is this the fault of the Architect who has overall responsibility for the adequacy of the design? Or is it the mechanical or electrical engineer? Should the Buyer take responsibility for the error in the design it gave to the Contractor? Is it the prime contractor's responsibility? Should the mechanical or electrical subcontractors have discovered the problem? The answer may be clear if the contract terms and conditions are all aligned and coordinated – for the designers and the contractors. But we know from experience that they are all too often not fully aligned and coordinated. The Owner prepares its own contract with the Architect or uses or adapts an industry standard one – probably after review with its lawyers. The Architect has its own form of contract that is uses with its sub-consultants. These contracts may not be the same for both sub-consultants if the relationships are different or if different people finalized the contracts. And the Contractor will have its own independent deals with its sub-contractors. These will not necessarily be the same. If the project were contracted using the traditional tender closing practice common in the United States and Canada, then the contractor may not even know what deal it really has with either of these companies. (We look at this crazy process in a bit more detail in Chapter 4).

There may be a marked disconnection between the contracts awarded on a single project.

2.6.1 Coordination

Coordination between contracts and contractors on a project is a major issue that is underrated on many projects and by many people in the business of contracting. First let us consider the issue of coordination between contracts in the control of the buyer, then we can consider co-ordination of contractors on a project.

Coordination of contracts is a two-pronged issue. We need to coordinate types of contracts and the buyer needs to co-ordinate the work of its contractors if there is more than one contractor on a project.

If contracts of different types are awarded on a project, the Buyer will inevitably end up with problems. For example, if one contractor has an incentive-based contract and another does not, the one with the incentive will push for priority and the other one will claim for delays that arise as a result of the first one getting the priority. If the priority is not given, then the first contractor will lodge a complaint or a claim based on the fact that the buyer got in the way of it achieving its target and earning its bonus. Similar anomalies routinely occur on projects where the buyer mixes and matches contracts without giving enough thought to the consequences of such an action.

Avoid mixing and matching contract terms and conditions without considering the impact of such combinations on the behaviors of the contractors you are retaining.

Another manifestation of this type of coordination is the use of an appropriate contract form to suit the situation. For example, it makes little sense to award a Stipulated Price Contract on a project where the specifications and design are incomplete for that work package. There will be so many Change Orders negotiated afterwards to address the guesses, changes and other alterations that it will effectively end up as a type of Cost Plus Contract. Similarly, awarding a Cost Plus or Unit Rate type of contract on a project where everything is clear and precisely defined serves little purpose as the contractor has enough information to accurately price the work and quote a firm price.

 Match the contract type to the situation.

The second issue is a simple and well-understood one. This is the business of coordinating the contractors you have retained for a project. It is not uncommon to include a clause that requires contractors to coordinate amongst themselves, but there are practical limitations to the effectiveness of this approach, so it needs to be supplemented with some overall management by the Owner or its representative.

 Without the active involvement of the Buyer, contractors who have to share space and resources or otherwise coordinate on a project are likely to run afoul of each other. This will cause delays and may lead to additional costs for the Buyer.

2.6.2 Management of Contracts and Contractors

Contracts do not manage themselves. We will see later that we need to pay attention to the detailed management of any contract. In addition to administration, there are a number of issues to be considered. Not least of these is the need to select a well thought out strategy for the contracts to be awarded with consideration for the following factors:

- The relationship between the contracting parties
- The amount of information that is available at the time a Contractor is selected
- The scope of the goods and services to be provided

- If co-operation between contractors and suppliers is needed, how this will be achieved
- The degree to which consistent incentives and bonus/penalty clauses will be used, and the implications of their use
- Timing of award of contracts
- Lead times required for the mobilization of contractors or the delivery of materials and equipment
- The role of each participant

Once the strategy is in place, the next step is to select the contractors and then to manage them. These steps are outlined in later chapters.

 Management of contractors does not just happen. It starts with a well thought out contracting strategy.

As soon as we have more than one contractor on a project, the contractors need to be managed and coordinated for the good of the buyer and the project. Generally, this is also good for the contractors if it is done properly.

2.7 Multiproject Contracting

To minimize lost time resulting from learning curves, understanding of client needs, and ways of working and to build relationships that enable more effective contracting, some buyers establish contractual relationships with a vendor that span more than one project. Let us look at a simple example:

A large technology company outsourced all of its manufacturing in the year 2000. Part of its confidence in doing so lay in its approach to management of its supply chain. It had built a number of symbiotic relationships over the years. It could rely on one particular company to manufacture extrusion molds for it. This relationship went to the point of the supplier's experts being involved with the design of new products; the reason for this was twofold: The first reason was the need to bring the supplier's expert opinion in mold design to bear on the detail of the parts it was designing in order to avoid manufacturing and quality problems and to minimize manufacturing costs. The second reason was that the buyer needed to nurture a working relationship with its suppliers so as to maintain a responsive and sustainable relationship. Time to market was critical. By effectively guaranteeing the supplier a steady workload, it was able to ensure priority for its own work, thus facilitating the overall process of bringing a new product to market quickly.

We see similar relationships between regular buyers of engineering and construction services. Here, the exchange is that the buyer gets a degree of preference, in terms of better unit rates and some priority in getting its work done in exchange for providing the engineering company with a base load of fairly steady work. The development of working relationships and processes that both companies understand is an added bonus, as this can lead to process improvement and a better product as a result . . . if managed well.

There are a number of reasons why a contract spanning more than one project may be of value to both the buyer and the vendor.

2.7.1 Evergreen Contracts

Evergreen contracts are usually signed for a specified period of time. During that time, the vendor undertakes to provide goods or services at an agreed set of unit rates. The purpose of such contracts can vary. From a buyer's perspective, this can include the following.

- Guaranteed rates for goods or services for a specified time
- Guaranteed access to the goods and services – especially if potential market shortages are foreseen
- Preferred rates in exchange for a guaranteed minimum purchase

Complementary benefits clearly exist for the vendor.

Evergreen contracts span any number of projects and take advantage of minimum guaranteed sales volumes to achieve both continuity of supply and cost savings.

2.7.2 Alliances

Alliances are a popular type of sustained relationship between a buyer and a vendor. The idea behind an Alliance is that there is some benefit to both sides. In the North American construction industry, these Alliances started to become a popular option in the early 1990s. The basic idea is that the buyer and the vendor work very closely together and share information more openly to the benefit of both parties. Part of the rationale is that it pays to avoid duplication of effort and this is easier to achieve if the two parties work more closely together.

Well-run Alliances benefit the Buyer through improved performance and reduced duplication of effort.

An effective alliance relationship needs to have value for both of the parties involved. Often the primary advantage for the supplier of goods and services is a base-load of work that supports other more aggressive business initiatives. This base load may be provided at lower-than-market pricing. Offset against this is a continued presence with the client and therefore increased opportunity for further work and a chance to build the relationships that encourage this. Also, there is a lower overhead, as the marketing and proposal preparation effort and cost are reduced or eliminated with such a "captive" client.

Vendors benefit from alliance relationships through reduced business development costs and a stable workload – often in exchange for lower fees.

The buyer, too, has potential advantages in such a relationship. First, the buyer gets to continue to work with a contractor or supplier who has been "trained" into understanding how they work and who has been sensitized to the special needs, politics, issues, and concerns of their organization. This pays dividends in the form of more effective briefing, reduced rework, and improved responsiveness to the specific needs and vagaries of the buyer organization.

Buyers benefit from alliance relationships through reduced costs and elimination of the learning curve required by a vendor providing goods and services for the first time. Also, a new team does not understand the buyer's operations as well as one that has experience in working with that customer.

If enough trust is built between the parties, additional synergies can evolve. For example, a true alliance does not have someone doing the work in the vendor part of the team and someone else redoing it in the buyer organization to make sure it was right! Business transactions are based to a certain extent on trust – invoicing is simplified with less backup being required and therefore with less checking effort to ensure that the backup is correct. All of this requires trust, and trust does not just happen. It is based on relationships, and these take time and effort to build. To be sustainable, trust requires a degree of stability within the team from both the vendor and the buyer. All too often, while the buyer insists on stability in the vendor part of the alliance, it is busy swapping its own people in and out of the team.

Moreover, in the numerous alleged alliances we have investigated as part of ongoing research, we see the underlying documentation and often the underlying attitudes of team members still firmly rooted in traditional contracting approaches. This naturally leads to behaviors that are the same as in conventional contracting and the advantages of alliances are essentially lost.

Despite the obvious benefits of eliminating duplication and the improved communication that should occur, many alliances are managed as more traditional contract relationships and the benefits do not materialize.

The very nature of an alliance is interesting. Whereas vendors typically feel they are able to provide a better service and are more comfortable in such an arrangement, the staff on the buyer's organization feel more exposed. This interesting observation came by accident while we were investigating quality management issues under different contracting strategies. Once we heard the story consistently, it became quite obvious – especially if we consider the normal human condition.

Under a traditional contract, the contractor is effectively "disposable" and can therefore be more easily assigned blame if something goes wrong. The traditional contracting approach accepts this common phenomenon. Alliances require collaboration and shared blame (perhaps "accountability" is a better word). The responses we got were quite consistent with this view of one of the key differences between traditional and more sophisticated business relationships. A blame culture in either alliance partner organization will likely mitigate against development of a sustainable and effective working relationship.

Sometimes the alliance group is treated as a separate company by the rest of the buyer's organization.

2.7.3 Term Contracts

Term contracts (e.g., annual supply contracts) are common for supply of materials and, sometimes, services such as maintenance, cleaning or temporary facilities (such as trailers and portable toilets). Construction contractors often sign up with suppliers of concrete, aggregates, rebar, drywall, lumber, and other construction material staples for a year (say), guaranteeing a minimum monthly consumption in exchange for preferred rates.

We see similar practices in larger organizations for purchase of car rentals, airline tickets, office supplies, cleaning or maintenance services and more.

Term contracts can include variations on leases for equipment, facilities and other products and services.

Term contracts exchange minimum sales for preferred rates.

In Chapter 2, we looked at some of the many options available to us for assembling contracting strategies. In Chapter 3, we look at what it takes to put the right pieces together to develop an effective contracting strategy.

Chapter 3

THOU SHALT LISTEN TO, AND UNDERSTAND, THE REAL WISHES AND NEEDS OF THE CUSTOMER

> *Contract strategies: intent and client objectives, contractor and supplier responses.*

The choices in developing a contracting strategy are almost boundless. To make a wise choice, a significant number of issues need to be considered. Which considerations are important are project – and circumstance – specific. To develop a realistic strategy, the particular intent and needs of the Client must be considered. We should take great care in making sure we really understand what the Client wants before we charge headlong into a recommendation for how to contract out what needs to be done or supplied.

3.1 Relevant Company Policies

Company policies can exist for many things. Usually, the larger the company or organization, the more policies there are. Some organizations – such as government agencies and others that handle large amounts of third-party (taxpayer, investor, policy holder, etc.) money – have a greater propensity to build extensive policies and are more likely to follow them too. Such policies will cover anything, from the use of electronic data interchange to contracting with employees' family businesses, spending authorities, and approval processes. Three of the most important policies in terms of influencing how we do business between companies are worth looking at. These three key policies are procurement, value for money and auditability.

Now that the client can't hear us or see us,
I hope he'll stop telling us what he wants.

3.1.1 Procurement Policy

Company policy is often steeped in tradition and protectionism. The latter is the result of many years of being audited and trying, long after the event, to justify an apparently inappropriate decision when the decision makers have moved on and the documentation is inadequate. The classic is: "Why did you not select the lowest bidder?" The fact that the company in question went into bankruptcy 4 weeks after the decision was made is not a factor. Nor is the fact that the project was completed under budget and ahead of schedule and that the work was well done! You did NOT pick the lowest bidder and, therefore, you paid too much for the work . . .

Many companies have a simple and arguably naïve approach to how to procure services. The simple answer is to tender the work and pick the cheapest vendor. This approach is time-honored and has a massive advantage to those in the buyer's organization who are involved: They will not get fired because they have done the right thing. But is it the right thing? We know that the lowest initial price and the lowest final cost are not necessarily the same thing and that the lowest initial cost can grow through changes, claims, and other ways.

 Competitive bidding – getting the lowest initial price – is an easy way to protect the decision maker. But it may cost the company more in the end.

Company policies also embed a number of other ill-conceived ideas based on making someone's job easier or making them look better. Here is another questionable, but common practice. We pay contractors late. The later, the better. The reason is that we are trying to maximize the use of operating capital and lines of credit. If however, access to this money costs us less than it does a contractor – and, with large companies, this is generally the case – then we are effectively forcing our suppliers to borrow money to cover our debt. This is of value only when the contractor is unaware of what we are doing, has a lower cost of borrowing, and is not at risk of being overextended. The possibility of having a competent contractor and meeting these criteria is pretty slim. Paying a contractor on time serves several purposes that are worth considering. One is that the contract terms are being adhered to. Another is that the contractor is more likely to provide service to the customers who pay on time.

Payment in 30 days or longer is often used to manage cash flow. This is only effective if the cost of borrowing is higher for the buyer than it is for the seller or if the seller's line of credit is larger than the buyer's. The trouble is that we generally do not know when these situations exist.

One reason for delayed payment is the need to obtain multiple approvals from different parts of the organization. This is most likely to happen when unusual purchases are made. We all like to think that our organization is efficient. If this is the case, then those items that we routinely procure are accommodated in internal procedures that have been streamlined. The essence of good payment processes lies in effective delegation. Senior management typically needs to sign onto spending the money on a project or program. Afterwards, they need to get involved only in cost overruns and at key decision points. If several departments or suborganizations are involved, they should also sign on at the outset, just like senior management. The details should be delegated to the individual or group best suited to the task. If the process of obtaining approvals exceeds the time available under the terms of your contract, either change the terms of the contract (if you can) or improve the approval process.

 Processing of progress payments may require a series of approvals that are related to different responsibilities in the organization. This may slow the process to unacceptable levels.

One specific case of payment delays is for changes made under the terms of the contract. These changes are often affected by company policies, and the result is a process that is out of line with the terms of most standard contracts and their conditions as used by that organization. The right approach is to bring these into line. Whether the change is made to policy or to the terms of the contract depends on whether the process is more important (change the contract terms) or the cost of doing business is more critical (change the process).

 Payment for changes and extras under the terms of a contract is often regulated by company policy. This policy should be reflected in the terms of the contract.

Change order processes are often dictated by a number of corporate policies. Two that constantly recur are limits on the value of change orders and the number of change orders that can be issued by the buyer's contract administrators. Another thing that constantly recurs is the way in which astute contract administrators who realize that there are several factors to balance, manage these limitations. Payment is important. But so is maintaining continuity of work, minimizing disruption, avoiding reworking items because approvals for a change were delayed, and more.

One common practice for dealing with, say, a spending approval limit for change orders of $5,000 when faced with a change order worth $14,000 is to split it into three change orders – each of which is less than the limit that can be approved by that person. Then there is the standard second limitation, which is to restrict the number of change orders as well. The common response to this is to rely on either hidden contingencies in the budget or to use another clause in the contract. In one case that I came across, change orders were paid for with boulders. Let me explain.

There was a limit on both the number of changes and the value of those changes that contract administrators could approve without having to go back to senior management. So they relied on a clause in their construction contract that allowed for payment, without any change orders, for the removal of boulders from the site. This was reasonable because history had shown that the normal site investigation could not determine how many boulders were hidden, and boulders cost more to remove than did regular dirt and often required overexcavation. The records also showed that there were many boulders on this Owner's sites. When the Owner found that it had to add a culvert here or modify a valve there or that a small bridge for a stream crossing had been missed, then the work was calculated and paid for in units of boulders. A culvert, for example, may be worth the same as the price for removing seventeen boulders. So the contract administrator and the contractor agreed on this exchange program and just got on with the work.

Before you berate the contract administrator for doing her employer wrong, consider the alternative. She would have had to prepare a change order and ask the contractor for a price. This priced change order would have required several approvals. The contract required that no work be done on a change before issue of the change order confirming agreement to do the work. The contractor could either continue doing the work originally mandated by the contract or stop and wait for approval for a change. If the former approach were used, a new change order would be required to add the cost of undoing this work and redoing it after installation of the culvert. If the latter, then a claim for delay would be in order and could probably be justified. The additional cost and the time delay associated with both of these options were saved by the contract administrator, albeit at the risk of being punished if she were found out. The lesson is simple: It often pays to be a little bolder in administering contracts in the best interests of all concerned.

Experienced contract administrators quickly discover the best ways to work with – and around – the policies when they see them getting in the way of good performance in the best interests of the organization.

3.1.2 Value for Money

Value for money is a requirement stated many times by many buyers as part of their procurement policy. The reality, as illustrated in Section 3.1.1, is not always in line with this statement. If value for money is a factor that is truly important, then a more sophisticated approach to setting contracting strategies is required. A full life-cycle view of the contract and what there is to support, needs to be considered and some interesting trade-offs are likely to come out of this review. The result should be the creation of real opportunities for more effective contracting. Other than the apparent initial price, some of the factors that should be considered include the following.

- Expertise of the Contractor
- Cost of managing the Vendor
- Quality of service and product
- Relevant experience of the Vendor
- The Vendor's team, how good they are and how easy they are to work with
- Track record of the Vendor in terms of disputes, reliability, financial stability, etc.

I would even consider talking to their subcontractors . . .

This is just for the selection of the Vendor. We need also to consider the work package that makes best sense, how multiple vendors are to be coordinated, how specific risks are managed, and how well we can define what is to be provided and by whom. These elements affect work packaging, choice of payment method, specific clauses in the contract, and how the contract should be administered. To be effective, all of these need to be understood and built into the contract at the outset.

 Policy should be steered toward obtaining value for money rather than the cheapest product. This provides a greater opportunity for more effective contracting.

Considering factors other than just the lowest initial cost requires decisions by the people setting the contracting strategy. These decisions need to be made with some sensitivity to the culture of the organization in which these decisions will be implemented. Decisions need to be justified. What is justifiable in one organization may be unacceptable in another. There is a good opportunity to establish the necessary ground rules for decisions in the process of determining the contracting strategy for a project or other venture. Properly documenting these decisions can save a lot of problems later on.

The ability to justify decisions is important in most organizations and in most boss-employee relationships. This often influences the way decisions are made and recorded.

Setting the strategy – and therefore the specific policy – for procurement on a project is important, as this can be used to clarify why decisions are made later. A risk-averse approach will lead to different decisions than will an approach in which risk is more readily accepted. For example, a low-risk strategy would include passing on a risk to a vendor rather than keeping it, regardless of the cost in hidden (and therefore unknown) premiums that the Contractor will inevitably charge. If the Buyer is willing to accept appropriate risk, it needs to assess the relative value of keeping or passing on that risk under the terms of the contract. A different set of contract terms and conditions will emerge.

Perceptions of what constitutes value for money vary depending on many factors, including company policy and traditions. We need to be sensitive to these.

3.1.3 Auditability

There is a need for any public enterprise or publicly traded company to be auditable. This means, in practical terms, that we need to be able to explain decisions. Policy and regulations are designed to help make decisions more a question of doing something acceptable, rather than doing the right thing. It is safer for management and those involved in any process to be supported in their work with easy and traceable decisions that do not assign any blame if they appear later to have been wrong decisions. Specifically, when it comes to managing spending of the organization's money on an external vendor, we need to take care that the rules are either followed or are managed in a traceable and supportable way. Our contracts should reflect this need.

The audit imperative is part of many businesses. Contracting processes need to reflect this need.

The process of selecting a Vendor needs to be supportable as leading to the right choice. The ground rules for selection of the best deal in the interests of the buyer's organization are important, because they are what the decision will be measured against. If this process is fair and appropriate, it also serves to show potential contractors that they will be selected using good, honest, and predictable process.

Auditability means that the selection of a contractor needs to be based on a process that can withstand the scrutiny of a third party.

If we now add the review of a third party (e.g., an auditor) to the mix, we are likely to see the reviewer biased by the credibility of the process that is under scrutiny. A big part of this "credibility," sadly, lies in compliance with policy or tradition.

The scrutiny of a third party is heavily influenced by the credibility of the process. One credible solution is that it is "company policy." Another variation of this is that "this is how we always do it."

3.2 Project Requirements

Clearly, the specific requirements of the project in question affect the way we put a contract strategy together. Surprisingly, the people who prepare the contract packages often poorly understand the specific requirements of the project. The connection between what is needed and the best way to procure it is commonly lost in the processes adopted by the organization charged with this function. This is often because key information is not shared with enough people involved in the processes that lead to establishing the most appropriate approach to contracting or else the approach is predetermined by habit, culture or corporate policy.

These requirements have technical, business, and social components to them. The technical components are usually well understood. The business components are often ignored, resulting in projects that have lower yields or returns or in contracts that go over budget. The social issues are also almost invariably ignored unless, like union jurisdictions or the clearly articulated needs of special interest groups, they directly affect the contract.

There is rarely any connection between the contracting strategy adopted for a project and what the requirements and drivers for the project are.

3.2.1 Balancing Performance, Cost, and Time

The driving factors behind a project that influence the approach to contracting are often not understood or else are not clearly articulated. These variables include contradictory elements such as urgency versus cost limits, scope certainty and quality expectations and the balance between these factors. On more than fifty projects, I have asked participants to prioritize project performance (defined as a blend of quality and scope), cost and schedule. Never, on any of these projects, have these factors had the same priorities in the minds of the key stakeholders. Without alignment of these factors, setting the right contracting strategy is virtually impossible.

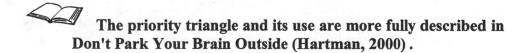

Use the priority triangle to understand and manage expectations around performance, cost, and schedule.

The priority triangle and its use are more fully described in Don't Park Your Brain Outside (Hartman, 2000).

3.2.2 Vendor Learning Curves

When we go shopping, we are affected by our experience in the store. This "buyer's experience" is a factor of increasing importance to many retailers as they fight for market share. This is as true for the experience in a real store as it is on the Internet. The experience with vendors on projects and in business generally is also important. Vendors can deliver more effectively for all concerned if they have learned their customer's business well enough to be truly responsive to the client's needs.

 For a vendor to be most effective, its staff needs to understand the needs, preferences and modus operandum of its client's organization.

It is not easy for a buyer to "train" its suppliers. Part of the reason for this is that often it is not a conscious process. The conscious part of establishing the working relationship is too frequently reduced to one in which the pecking order is established, rather than one that fosters trying to get the best out of the business partner. When the buyer's organization recognizes that it needs to train its suppliers, it usually recognizes that minimizing this effort leads to better results earlier. One effective way to minimize the effort is to use vendors that are already trained.

The process of "training" a vendor in the various aspects of project delivery as expected by the client is an avoidable overhead in terms of both cost and time.

Vendors should be motivated to understand the needs of their customers. The better they understand them, the more likely they are to be able to respond, win more work, get paid and establish the relationships necessary for everyone to win from the collaboration.

 From a vendor's perspective, understanding the client's needs and being responsive to them is a major advantage.

Now that we have established that there is an advantage to knowing and understanding the client, there is also a potential disadvantage. If we know too much, we will price a proposal based on what we know. This may well lead to a price that is higher than that of our competition. We will have included for all of those elements that we know about and that the competition cannot know because they have not worked on this project or with this client in the past.

In my contractor days, an earthworks subcontractor once told me that he would drive past any prospective site to determine what was involved. The more he needed the work, the faster he would drive. That way the site not only appeared smaller but he saw fewer problems. He would be much more competitive as a result! Although this was said in jest, there is a depressing reality to this. I do not recall ever having won a competitive tender after we had been asked by the owner or the designer to review the drawings (for constructability or feasibility or to offer suggestions for cost savings, for example). We knew too much and priced the risks and problems that we might have missed had we been just another bidder.

 Knowing too much about a project can lead to losing work to another contractor that has not included in its price for all the things we know about.

Just five drawings but seven volumes of specifications
– must be a very worried client.

3.2.3 Reputation and Relationships

We have seen that we can know too much for our own good. Similarly, the reputation of a client or a vendor can affect the way we are prepared to do business with them and, hence, our contracting strategy.

For many buyers of goods and services, there is more to buying than just the lowest price. Most buyers also need to be reassured that they are actually going to get what they want and that it will be delivered in a reasonable time frame and with enough care and attention that the process of acquisition is not too cumbersome. A big part of achieving this is to work with vendors that are reliable and consistent in their performance. Often, to address this need, owners will consider inviting only those vendors that they feel will meet these performance criteria. This is why, on some projects, invited bids will be considered.

 The reliability of a vendor is important to most clients.

Selection of a vendor from an invited list is often based on price. The assumption is that all of the invited bidders will deliver a similar level of service and a similar quality of end product. That is why they were prequalified and invited to make an offer to do the work. Beyond this, if price is not the only consideration, then other factors need to be taken into account. These factors are frequently assessed in a way that allows for comparison of the bids in ways beyond just price. One example of what has been done is the use of owner costs associated with expected contractor performance. Consider the following situation in which just two prequalified contractors are being considered.

Assessment Component	Contractor A	Contractor B
Base bid	**$10,000,000**	**$9,700,000**
Estimated administrative cost	$320,000	$535,000
Estimated cost of managing claims and changes	$25,000	$150,000
Allowance for changes and claims settlement	$ 50,000	$250,000
Estimated final cost	**$10,345,000**	**$10,635,000**

Table 3-1. Bid assessment with risk management.

In this example, if the Owner considers only the base bid, it will pick contractor B. If, however, it has considerable experience with both contractors, then it may consider their track record and may assemble a table that adds other items to the selection equation. If the result were as shown in Table 3-1, it would be wise to select contractor A. Justifying such a decision sometimes needs third-party support. If the reputation of one contractor compared with another is validated by reference through other clients or others that have worked with the Contractor, then it is easier to make and justify the better decision.

It is easier to justify selection of a vendor other than the lowest bidder if it has a good reputation and if existing relationships already work well.

The relationships between people will affect how we do business. Many of us feel it is better to do business with people we trust. Sometimes we are happier with a business partner we know and maybe do not fully trust, than with a complete stranger. Another advantage of doing business with a known partner is that we know its habits and traits, we know how it operates and we know how to get things done through the other's organization. Whether we are a buyer trying to encourage a vendor to look after our interests and deliver a quality product efficiently or we are a vendor trying to understand the customer's needs and get paid for meeting them, the association and the improved communication that results from a good working relationship will benefit all involved.

From both a buyer and a seller perspective, doing business with a known partner is usually easier and has a lower risk than working with a new and untried relationship.

3.2.4 Clarity of Requirements and Specifications

Most practitioners, in managing contracts, will have stumbled across the classic vague specifications that have plagued contractors forever. Here are some real examples.

"[The quality] is to be to the satisfaction of the Engineer."

"The brick shall be a mix of three colors as specified elsewhere and shall be blended and placed in the wall in a manner that is acceptable to the Architect."

"The system shall perform better than the current system."

"The hardware shall be configured to accommodate future expansions and changes."

"The contractor is deemed to have included for everything both shown and not shown on the drawings."

These are some of the more obvious examples of things that are essentially based on an intangible and that a sane contractor does not try to price.

We look at this topic in more detail in Chapter 5.

There is too much guesswork for anyone's comfort when specifications are vague, ambiguous, or incomplete.

Rather than trying to capture vague and incomplete ideas in a contract, it pays to acknowledge what we do not know yet – perhaps because the design is incomplete – and structure the contracting strategy to reflect this. The reason this is a wise option is that vendors tend to price unknowns conservatively and may even place conditions on the price. Alternatively, vendors may rely on the tendency for the courts to rule in favor of the party that interpreted a vague or ambiguous term of the contract over the party that was responsible for writing it. Either way, the buyer loses if the strategy is inappropriate.

A good contracting strategy should reflect the amount of definition of the end product that is possible at the time a contractor is selected.

3.2.5 Business Drivers for the Project

Let us consider the business drivers for a particular project. The buyer may embark on a project to construct a warehouse. What the buyer wants or needs may vary significantly depending on the business it is in.

If the buyer is a real estate developer, it may wish to own the facility and lease all or some of the space to tenants. In this case, it may need to borrow money from a bank to finance the construction using the land as collateral against a project construction loan. After construction, this loan would then be converted to a mortgage that is approved based on revenues from a tenant. If one tenant is being sought, the owner of the land may prefer a design/build approach to attract a tenant that wants a tailored facility to suit its needs. If it is a speculative building for a landowner who does not normally build warehouses, then a deal that offers a turnkey solution including project financing may appeal.

If the client owns the land and is expanding an existing operation, it may consider a lease-back arrangement to preserve its operating capital for expansion of the business rather than construction of the warehouse. Under such an arrangement it could assign the land to the builder who then builds the facility, finding the money from, say, a pension fund or a large insurance company. Once construction is complete, the contractor transfers title to the financing organization that then holds the property for the life of the lease. At the end of the lease, the original owner of the land can purchase the land and the improvement (the warehouse) for a nominal amount.

If the warehouse is for a large organization, it may wish to select a designer, complete the design and then tender the work to a number of contractors that are competent to build it.

Perhaps the owner builds many of these warehouses and has an alliance agreement with a design firm and a contractor. In such a situation, it may prefer to negotiate the terms for the construction of the building with its alliance partners.

 The business drivers for the project could affect a number of contracting strategy decisions.

In the previous examples, we can see that the there are a number of ways of packaging work to meet the specific business circumstances of a particular project and the needs of the buyer. The need for financing, whether the financing is off balance sheet or not, speed to completion, whom we already do business with and many other factors can influence the way in which we structure our approach to contracting – even for something as simple as building a warehouse.

How the work is packaged may be influenced by financing constraints, existing business relationships, longer-term contracts, urgency of delivery, and market conditions.

3.3   On-Staff Expertise and Availability of Key Skills and People

One of the main reasons for contracting out work or supply of materials or equipment is the simple fact that we do not have the resources or the skills in-house to deliver what is required. This clearly plays a big role in developing the decision to outsource anything. It is, therefore, fundamental to developing the right contracting strategy. When we buy some everyday item, we are usually fairly comfortable with the process. We know what we want and can select a product that meets our needs. Other factors,such as price versus expediency, come into the decision, as do factors such as ease of use and durability. Even for trivial purchases, such as buying a stapler, we develop a strategy for buying it. Maybe we decide that convenience is the prime driver and we pick one up on the way home. Maybe we want to replace the one we broke today with an identical one, so we prepare to try more than one source in case one does not have it. Alternatively, we may be in no rush and we look out for a sale . . .

The stapler example is easy. We know what it is and where to buy one. We know what our priorities are. With larger acquisitions, we move further away from complete knowledge and rely instead on others' expertise and the decision process of a business or a team. This is when knowing what we know and what we do not know becomes important. If we are buying something that we need but do not understand well, we may seek others' expertise to provide some independent advice and guidance.

The buyer's need for expertise cannot be totally replaced by retaining a contractor.

Specific in-house expertise is required to manage contractors and suppliers. How this is achieved may be through retention of a third-party professional in the absence of a knowledgeable sponsor or contracts manager from within the buyer's organization.

As a vendor, we need to be sensitive to the skills that exist in our customer's organization. A lack of experience and skill in the area of what we are providing is often associated with unrealistic expectations. Also, any shortage of expertise in a client organization is often replaced with a commensurate amount (or more) of anxiety. This anxiety will manifest itself in many forms and will therefore need to be addressed.

From another perspective, vendors also need to assess their available resources and expertise, as this will help formulate the contracting strategy they needs to adopt to meet obligations under the contract with their customer. It is not unusual for a vendor to offer more than what it can do with its in-house expertise or resources. The solution is well established. The work is subcontracted, and materials and equipment are purchased from others. Again, the strategy can vary considerably from one vendor to another. Some like to work with preferred partners, whereas others competitively tender anything.

As a vendor, we need to consider availability of resources and the extent to which we may need to outsource through subcontracting parts of the scope of the work.

3.3.1 Technical and Other Complexities

If life were really simple and everything that was needed could be supplied by just about anyone, contracting strategies may be a lot easier to put together. This is not the case. Sometimes there is only one reliable source for what we want. At other times, market conditions restrict the number of sources for what we want to a very few. And all too often, these few are the ones that are not as good as those who are already really busy. Now we need to consider the technical and organizational complexities of what we are procuring. It may be that we need the expertise of several vendors to assemble what we need. This is common. Also common is the need to keep business or other operations going while the project we are procuring is assembled around us. Examples of this are a building renovation and upgrade to a newer control system for manufacturing, improved software for a specific application, or retooling of a manufacturing line. Any of these – and innumerable other – examples are complex for a number of reasons. First, there may be technical challenges in integrating new and older technologies. There may well be surprises in what exists. It is not uncommon to find someone has modified a facility or a system without telling anyone. If we also have to keep running our business while the procured changes take place around us, scheduling and other special requirements (temporary relocations, rental of replacement equipment, and other transition steps) need to be considered in determining the type of contracting arrangements that we need to meet the specific requirements of the project. If production is going to be affected, or if part of a process plant or hospital needs to be shut down for a while, management needs to be fully involved and informed so that the best arrangements for the staff and for the buyer's customers can be made.

Before finalizing a contracting strategy, we need to consider the complexities of the work being done and their implications on the management process.

Conventionally we refer to complexity as a way of describing how intricate a project is: technologies, departments involved, number of contractors, and so on. This is fair. But if we look a bit deeper, we can see that this complexity really affects our ability to communicate clearly and without ambiguity or hidden agendas more than it affects any other facet. We can coordinate effectively if we communicate well. We can identify and address problems early. We can manage situations better if we all know what is going on.

When we see a project that is referred to as complex, beware primarily of the communication needs that this creates. One common breakdown in communication is that vendors and buyers alike assume that the other party knows their business. How many times have we heard: "Well, you should have known that we ALWAYS . . . [fill in the blank]." As we increase the number of technical, business, and cultural interfaces between participants in a project, we also need to increase the amount of care that is needed in managing communication and coordination.

Complexity is more closely related to the number of technical, business, and cultural differences that need to be overcome than to the technologies and business issues themselves.

The greater the complexity, the more we need to plan for communication and coordination challenges between the various contracts and contractors.

3.3.2 Project Priority in Sponsor Organization

A million-dollar contract for a Fortune 100 company is probably no big deal, unless its outcome will have a significant effect on the performance of the company. A million-dollar contract awarded to, or by, a small family business is a huge issue. The latter is much more likely to get attention from management and the owners than is the same-size project in a large organization.

> **The bigger the contract, the more likely that there will be attention from management. This can be good and bad.**

As the relative importance of a contract grows, so the need to manage the expectations and anxieties of the customer's management team will grow. The development of a sound contracting strategy, buy-in from management to this strategy, and its communication to all stakeholders are of greater importance if the contract being planned for is likely to have a big effect on the buyer's organization. The need for an appropriate contracting strategy grows in these circumstances for both the owner and the contractors involved.

> **The more important the project and the higher the profile of the contract, the more difficult it is to manage expectations of stakeholders. Effective communication of the planned contracting strategy and the associated risks becomes increasingly important.**

3.3.3 Impact of Funding for the Project

If the money to pay for the contract is in the bank, waiting to be spent, then funding will not directly affect the terms and conditions of the contract. This is not often the case, so we need to consider the likely effect on the contracting strategy of funding from sources other than "ready cash".

If the buyer is funding the project from revenues or is releasing money from investments based on an expected cash demand from the contractor, then this should be reflected in the contract. Either the progress payments are tailored to the availability of the buyer's cash or the contractor should be asked for a cash flow projection for progress payments. This may be linked to payment methods (e.g., weekly versus monthly; 30 days to pay or 60; method of measurement for progress payments) defined in the contract.

As a contractor, it pays to understand the implications of the payment terms proposed or dictated in the contract document. These affect the amount of work that the vendor has to finance on behalf of the buyer. And this, in turn, affects the pricing of the contract.

Lenders and bankers may also have special requirements related to the performance of contractors on projects on which they would be exposed if the project were to fail (see Not a Chip off the Old Block).

Not a Chip off the Old Block

A lumber business decided to branch out. It wanted to use some of its waste product – and that of its competition – to manufacture chipboard. It commissioned a contractor to do some preliminary work. The contractor provided design, procurement, and construction services. It developed a plan and a budget of $100 million for the plant. It was prepared to enter into a fixed price contract for this amount to complete the design and build the factory.

The owner needed to finance the project and went to a bank with a business plan and a copy of the contract that it would sign with the contractor if the project were to go ahead. The bank was concerned that the plant would work and that the buyer could operate it at full production, so it insisted that a clause be added to the contract to this effect as a condition of financing the project.

The contractor was not going to operate the plant, so it saw the clause as adding considerable risk in the event that the owner's operations staff failed to perform. The contractor was confident of its design and construction capabilities, but the operators would be new to the buyer's organization and were essentially untested. As a result, the contractor modified its offer: It added another $10 million to cover the risk. The project went ahead at $110 million.

The source of funding can influence the contracting strategy and may affect the terms and conditions for payment.

3.3.4 Politics and Compliance Issues

"Politics" spans a broad range of topics from use of local companies to long-term alliances and from risk tolerance of the buyer corporation and the vendors to corporate culture, turf issues, and power struggles. Many of these doubtless affect the day-to-day work we do. The important elements that may affect contracting strategy in an intelligent way are those that relate to the nature of our business dealings with organizations outside our own. Specific elements to consider include the following.

- The effect of local or national requirements for "local content" – to what extent do businesses exist in the physical area in which the contract will operate
- Requirements for technology transfer
- Compliance with laws, codes of practice, and customs
- Assignment of responsibility between parties
- Attitudes to the significance of a contract (For example, is it a summary of the business deal, or is it the starting point for negotiation?)
- Policy toward paying – or receiving – commissions, gratuities, "special bonuses" or other ex gratia payments
- Use of existing relationships with others who may be involved in the deal embodied in the contract
- Willingness to work with specific other organizations.

This is not an exhaustive list. It merely serves to identify the need to look out for potential issues that will nip at our heels and cause us grief if not carefully considered when appropriate. Typically, the further we are from our home base, the more likely that this list will be appropriate, and the greater the probability that we will need to add more to the list.

Politics may also play a role in molding the strategy adopted for a project. It is not uncommon to see a certain amount of local content either required or expected in a project.

Longer-term relationships with preferred vendors may also dictate (or strongly encourage) their use on a project.

In Sections 3.1 through 3.3, we considered issues that primarily reflect the buyer's perspective. What are the equivalent concerns as seen by the vendor? This question is too rarely considered by owners and their representatives in the contracting process.

3.4 Risks Associated with the Client

As a smart business practice, many organizations routinely investigate the capability of the client to pay for the goods or services that are to be provided before a contract is finalized. This due diligence pays for itself, as often does care in understanding the knowledge and capability of the client organization, their approach or attitude to vendors generally, and how fair or understanding they are in the management of their procurement and their contracts.

The effort taken in understanding a prospective client before dealing with it is invariably worthwhile.

3.4.1 Can the Client Pay?

If we do work for a customer, we expect to be paid for the effort, as well as for the opportunity. The greater the risk, the greater our expectation of return on our investment should be. A critical part of the process of qualifying a customer is an assessment of whether we will be able to recover our costs for doing the work.

There are many factors that will affect our ability as a vendor to collect payment. Usually the process is easiest if the funding is guaranteed in some way. In private industry, this often is not the case, and we need to rely on the contract alone. The next best case is that the money is budgeted and sourced within the buyer's organization. The sourcing of money is important, as it may be tied to a range of other factors that the contract administrator does not control. Some owners use cash flow derived from revenues. Others need to manage their treasury to release funds from short-term investments or other sources. Others have negotiated a loan or an extended line of credit. Yet others use nonrecourse financing, which typically brings greater demands from the lenders because of the increased risk exposure. Yet others may rely on annual budgets. If a contract spans more than one year, the budget for anything after the first year may not be secured.

It pays to understand how the customer plans to obtain the funds for your contract. It also pays to understand if the client is aware of the risks associated with the contract and what this means in terms of carrying an adequate budget. Poorly defined or volatile work will tend to increase the price of the contract over time, and the inevitable changes and their effects are assessed and added in to the original quote. If the owner carries too little contingency, then it is likely to have problems in meeting later payments. Or these payments may be delayed if approvals are required at a high level (such as the board of directors).

To mitigate some of the risk from the buyer's perspective, clauses may be added to the contract to address these potential problems. Some may add considerable risk to the vendor. Others may allow for payment deferrals.

The source of financing may influence the specific terms and conditions of the contract.

What the client has allowed in its budget and contingencies will influence the way it can or will respond to additional costs should they occur in the contract.

It is quite common for a buyer to include some safeguards to protect it from abuse by the vendor during the life of the contract. Holding back part of the payment is mandatory in many jurisdictions. Even when it is not, some buyers have a policy of holding back on payments to cover for potential deficiencies and problems in completing the contract. Others may have additional steps in the approval of progress payments included in their contract terms and conditions.

Most experienced lenders and investors have specific requirements that will affect the administration of the contract. These modifications can affect the terms and conditions of the contract (see Not a Chip off the Old Block). Often this means additional risk is passed to the buyer, who then may pass it to the vendor, or the buyer may incorporate the risk directly into the vendor's contract as a condition of the loan.

Based on this study, we need to reduce the cost by $1.49 to make this project succeed.

Investors and lenders may require specific safeguards in the contract and in the progress payment process to ensure that their exposure is minimized.

Another example of potential payment problems is the use of funds that carry temporal or other limitations. Experienced buyers plan for their expenditures. Such planning usually involves a chain of people with different responsibilities and priorities. Planning for release of funds to meet contract commitments involves the corporation's treasury function, accounting, the contract administrator, and whoever is responsible for providing an estimate of the cash flow requirements. Not only does this chain frequently miscommunicate, but errors in estimating, calculation, and comprehension may creep in at any point in the chain. Add to this the effect of other decisions (such as a new contractual commitment) that are external to this process as well as the effect of, for example, reduced margins in a suddenly depressed market, and we have the potential to not get paid fully or on time. Some due diligence in selecting our customers is usually a good idea.

It was once pointed out to me by a CFO of a construction company I was managing that for every dollar we lost we would have to complete a new contract for at least $50 worth of construction just to get back to where we were before we lost that dollar.

 To what extent does the owner depend on cash flow to finance this project? The constraints may be created by budget cycles, client business success, or credit limits.

Although perhaps not a sure way of flushing out these types of issues, it pays to read between the lines on payment terms and the ways of not paying that a buyer has included in its contract. These can sometimes provide useful indicators of potential constraints that the buyer faces or anticipates in being able to make payments to its vendors. I generally look for what appear to be arguably inappropriate clauses. Examples include the owner's right to stop the work at any time and indefinitely, to postpone or cancel parts of the work at any time and for any reason and to be the final arbiter on any decisions relating to payment, changes or extras to the contract.

The appropriateness of contract and risk allocation can point to potential future problems with payment.

A fairly common clause in construction contracts is one that requires full agreement of all parties to the terms and conditions related to a change before any work is done on the project. Interestingly, common practice in construction is to start the work on a change despite existence of this clause.

The right to payment for changes may be lost under some contract conditions if the work is started before the change order is formally issued.

Clients who are buying a significant item, and who do not routinely acquire such an item, may have unrealistic expectations of their contractors. A client, for example, may wish to build a new office building. The last time it did so was more than 20 years earlier, and none of the people involved with that project are with the organization any more. The client is not in the construction business. What does it do? Conventional thought is to retain an architect and follow the traditional process of design, tender for construction, and build based on a stipulated price contract. In this situation, a buyer cannot be blamed for thinking that it has the final price when the construction contract is signed. As people in the construction business know, design changes and adjustments, differences in site conditions, field changes and more will more than likely lead to alterations in both the fee charged by the designers and the final price for construction and commissioning of the building.

Clients who have little or no experience with this type of project may have misaligned expectations, which will lead to difficulty in its Contractors recovering payment to which it is entitled.

3.4.2 Expertise and Experience of the Client

The greater the expertise and experience of the client, the more likely that it will have the corporate capacity to manage more complex approaches to the procurement process. The greater the in-house capacity for management and coordination, the more likely the buyer is to manage the different aspects of a project by not vertically integrating its contracts. The converse holds true: Design/build approaches appeal to the one-stop shopper.

It is easier for vendors to work with experienced and expert buyer organizations and individuals, as they understand the complexities and do not have to be "trained."

☺   Regular buyers of similar goods and services understand industry practices better than novices or infrequent buyers. If the contract and associated processes are complex, then working with experienced owner representatives is easier because they know what to expect.

☺   The customer's expertise in any relevant technology will help a contractor in that the specifications and client guidance will be more accurate and focused.

☺   Client expertise in management of similar contracts will make management of its expectations easier.

Do we really want to contract with a buyer organization that has a reputation for being litigious? Are we prepared to wait for payment from a company that is famous for its "we-pay-in-180-days-whenever-we-can" policy? Is it worth trying to do business with an organization that has a well-earned reputation for being difficult to deal with – and that most of the competition will not do any work for? These look like much easier questions than they really are. We live in a competitive world, and opportunities sometimes lie where others fear to tread. Sometimes clients change. Other times we find a way to work with them so everyone wins. In the end, the key is to understand our business well enough to know whether a particular opportunity is worth pursuing.

It pays to check out and carefully consider the track record of the buyer as a client (i.e., its reputation).

3.4.3 Client's Attitude to Contractors and Suppliers

We touched a bit on the attitudes of buyers in Section 3.4.2. Most of us, as vendors, value a peer relationship with our clients. We know that this works better because it enables communication and a degree of give-and-take that simply makes projects more successful. We are able to build the relationships that enable greater cooperation, out of which grows cost-effectiveness.

We are, however, schizophrenic. While we understand the peer relationship as a Vendor, we struggle with it as a Buyer. When we are a Buyer, it is our money. We know what we want, and the Vendor is a servant to our role as master. The master-servant relationship therefore dominates because of the golden rule: Whoever has the gold makes the rules.

 Master-servant relationships still seem to dominate the contracting scene.

A popular relationship in contracting since the mid-1990s has been the Partnership and the Alliance. The idea was that sustained relationships develop between the parties that contract with each other. Several things seem to happen when these relationships are developed. The most common tendency in the alliance relationships that we studied was for them to be little more than the same old contract with a few fancy words laid over them. Certainly it was rare to see the changes in behavior and attitude that were necessary for the alliance to flourish. In true partnership relations, one partner can make decisions that bind all the other partners (that is, after all, one of the legal principles of such a constitution). Perhaps more important than this is the sense behind it: We are in this together. Arriving at such a relationship takes time and requires a high level of trust. Most of us are simply not built that way.

 True "partners" in partnerships and alliances are quite rare.

Moving from the adversarial relationship that is so common in many contracts to one of collaboration is difficult. However, when this happens, the business experience changes significantly:

- The parties identify with each other more readily
- Communication is more open because there is a much lower threat of recrimination or reprisal
- More and better information is exchanged
- The parties support each other and understand and align their objectives
- "Your" problem becomes "our" problem.

This partial list serves to identify just some of the benefits observed in the most effective partnerships.

A cooperative rather than an adversarial approach by both parties can lead to significantly better results for all.

There are a significant number of buyer organizations that have rigid and immovable policies around purchasing and contracting. These inevitably restrict the options that contract administrators in those organizations have for selecting – or even having – a contracting strategy other than "that is how we do it here." Another outcome of policy dominating discretion on the part of the person charged with making contracts work is that many decisions are pre-made based on the organization's policy. It pays to understand this, because if we know the policy, we can use it to our advantage (see Only One-Third).

In some organizations, policy-bound administrators rather than contract managers dominate. In this situation, it pays to know the policy!

3.4.4 Fairness in Contract Administration

Fairness of the approach taken by the owner may be managed by the contractors who identify a pattern of behavior. Recognition of this type of pattern is not uncommon. In my days as a consulting engineer, I was often asked by contractors who were invited to submit a tender, who the Client's project manager or contracts administrator was. In an earlier book, I told the story of a person who routinely attracted a 5% premium in quotes from contractors. The feedback I got on this story suggests that this is far more common than I originally thought. In my days as a contractor, I recall subcontractors wanting to know who the project manager and the superintendent were before they gave us a final price on a project.

A reputation for fairness in managing contracts brings the initial price down. Whether the final cost is lower or not does not appear to have been studied and may prove to be impossible to assess. What we can glean from observation is that the way contractors price their work is based, in some part, on the perception of the risk associated with the owner's approach to management and administration of the contract after award.

Only One-Third

This story comes from the administrator responsible for a remarkable approach to managing claims and extras on contracts. The project was large and had a very visible end date, linked to an international event. The project manager was a prominent local businessman brought in to "fix" the project because it was seen to be running late and over budget. The contract administrator was connected to this new project manager. He was presenting at a local professional association meeting on the management of construction claims.

His process was easy. All he did was take one-third of any claim and then fight tooth and nail to make sure that the settlement was no more than that amount. He was a hero in the eyes of his boss, who saw two-thirds of the amounts being claimed for changes and extras disappearing from his overruns.

It is hardly worth pointing out that the contractors understood this approach very quickly. They simply tripled – or more – the value of change orders, claims, requests for payment for extras, and any other priced item that came under this person's assessment. The trick was not to make it obvious, so some brinkmanship in presenting the claim and threatening legal action was considered essential to the process. In the end, the contractors all settled for only one-third of the payments that they requested, walking away with all that they needed.

There is an old saying in the construction industry: *"All contractors were born at night – but it wasn't **last** night!"*

It is generally not a bad guide to consider the decision-making (money spending?) authority of the person we are dealing with. The greater the authority, the easier it is to work with that person, as they will have a shorter command chain to manage when addressing contract changes. Also, the greater the spending authority, usually the greater the power and influence of that person.

 The greater the decision-making authority of the people involved, the easier it is to manage and administer the contract.

Creative administrators know how to work around the limitations imposed on them. There are endless examples of how people manage their work environment. Sometimes this type of anarchy is good for the organization as it gets things done. At other times it is bad, as it muddies the water and reduces effective control of the business by management. The story in Section 3.1.1 about how one person used boulders to pay for changes is just one manifestation of human creativity in the face of bureaucracy.

 The spending limit is a barrier to expeditious processing of changes, unless the administrator is creative!

Please do not get the wrong impression here. I am not advocating that anyone plays games with the systems and processes that the organization they work for has put in place. Challenge them, certainly. Know how to manage effectively in the best interests of your organization with minimum effect on the systems that appear to get in the way. This said, sometimes we need to change or improve the way we do business to accommodate changes that occur. It also pays to understand how a client operates within its own systems and how creative the contract administrator may be while working safely and ethically within this environment and its limitations.

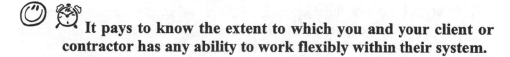 **It pays to know the extent to which you and your client or contractor has any ability to work flexibly within their system.**

One surprisingly common challenge in managing contracts is a fundamental difference in the opinion of one party to the contract versus the other as to what a contract clause really means. This came out in some research we did into how people interpret clauses and assign risks. This is explored in more detail in Chapter 5. For the context of Chapter 3, it is important to note that anything we can do to obtain a common understanding of the intent of a contract <u>before</u> we commit to it is invariably worthwhile.

Get a feel for the contract wording and how it is interpreted by the other party.

Contractors are just a part of the client's team. As such, we need to understand the relationships that the contractor has with others on the team. For example, if the client has a close and long-established relationship with another member of the team, is it likely to favor the other party if there is conflict between us and that team member? Similarly, would we have an advantage if our relationship is stronger than that of another supplier or contractor on the Client's project team?

It helps to understand the Client's relationship with other parties (suppliers, other contractors, consultants) as these can affect how our contract will be administered.

The tender documents are usually focused on what the Contractor must do. Sometimes this is to the exclusion of some of the critical interfaces that need to be managed in delivery of the Contractor's obligations under the contract. For example, there may be other contractors or suppliers of critical information or equipment who need to be closely coordinated with for the effective execution of the work. These other entities may not all be clearly identified in the contract. Furthermore, the buyer's contracts with these others may not provide for the coordination effort required for each of us to meet our obligations effectively. This needs to be checked out and assessed in terms of risk, and management of the exposure, that such interfaces may create. Some of the typical interfaces we need to consider are listed below:

- Regulatory approvals and permits
- Acquisition of access to the location of the work
- Design details

- Construction approvals
- Provision of temporary services
- Equipment and materials provided by the owner, directly or indirectly
- Specialist input from consultants
- Actions and work being done directly by the buyer
- Other contractors performing work before, concurrently with, or after our work
- Stage gates or other processes that may lead to work stoppages or changes.

These matters and the associated processes for managing them need to be identified or addressed at the outset, before any firm price is submitted as a bid. In the absence of any clarity or ability to resolve these issues before submitting an offer to do the work, it pays to qualify or clarify the assumptions that have been made in addressing these interfaces.

The roles of all related parties who we need to coordinate with, such as other contractors, designers and suppliers, need to be clearly articulated early in the tendering process – especially if this is not clear in the tender documents.

Developing an effective contracting strategy is a critical part of the process of outsourcing work on a project. We need to consider the specific circumstances of the project, as well as the nature of the work being done. The traditions and policies of the companies involved, especially the buyer, should be accommodated – or formally challenged if needed – in the development of the most appropriate approach to selection of a contractor and how the work is to be assigned to the contractors if more than one is involved.

Chapter 4

THOU SHALT NOT BLINDLY PICK THE CONTRACTOR WHO WAS CHEAPEST BECAUSE IT MAY HAVE MADE THE BIGGEST MISTAKE

> *Contractor or supplier selection: picking partners for success.*

The way in which we pick our partners often sets the tone for the rest of the contract relationship. Whether we use the same contractors and suppliers time and again or we bid competitively in the open market each time, we need to understand the effect of the bid and award process on the rest of the contract life.

4.1 Picking a Contractor: Selection Strategy

Think about how you pick your contractors and suppliers. Is the approach based fundamentally on company or organizational policy? If so, then how responsive is your approach to the specific needs and circumstances of an acquisition of goods or services? In one company that I took over, we had a standard way of selecting subcontractors and suppliers – and we used the same approach regardless of the competitive situation we were in. The result, predictably, was that we were no more competitive than all the other contractors that were using the same suppliers for the same projects. What this meant was that we relied more on luck than anything else to obtain any edge in our bids to our clients. This is not a logical way of doing business. We need to be more flexible in how we go about engaging the help we need to deliver a project.

No cheating now, we are selecting our contractor.

 Many selection strategies are essentially based on "this is how we do it" rather than on a carefully thought-out approach.

Instead of just doing what we have always done to award a contract for goods or services, we should consider deliberately planning the best approach to contracting given the unique circumstances of the situation, market conditions and the project we are working on. In a very busy market, we may have little choice with whom we may work. New entries into the business may offer a better price but may not have either the expertise or the financial capacity to handle what we have in mind.

The obvious response to this situation seems to be to pick a more experienced or at least a more financially stable supplier or contractor. But the question then, becomes: Are we doing ourselves a disservice?

Conversely, our policy may be to retain the contractor with the lowest price, regardless of other circumstances. Now, the question becomes, are we taking on a greater liability than if we went with a contractor that offered a more realistic quote? Perhaps the additional cost of doing so will be less than the cost of dealing with and managing the "cheaper" contractor over the course of the project for which it is being retained. Other considerations include the likely cost of managing one contractor over the other and the likely effect of inadequate or late performance on the organization or other suppliers. Clearly, a more complete assessment of a given situation will provide insights to the best approach to contractor selection. More important is the need for such an analysis and the absence of a single "correct" solution that is appropriate to all situations and circumstances.

A deliberately thought-out and planned strategy to contracting out work will save time, money and trouble in almost every case.

Wherever we are in the food chain, we need to consider the likely outcome of a particular selection process for our suppliers. Competitive tendering has a role, but it is not the only way to reach a competitive position at the end of the day. A good starting point is to look at our customer and to understand the implications of a particular contractor selection process on the relationship we have with that client. What does this mean?

If we are competing solely on price and if the scope and specifications of the work to be done are completely and accurately defined, then the opportunity for innovation is limited to process and methods for our work. If our work involves participation with subcontractors, then coordination of this work can yield dividends. The ability to coordinate with subcontractors depends on a number of factors too. How busy are they? What has our relationship with them been in the past? Can we trust them to protect our innovative ideas for creating an edge on this project? What about the other way round? Which combination of contractors and suppliers works – and which ones are antagonistic to each other and will generate delays and other problems on the project later on?

 Even for a contractor, subcontracting strategies need to be thought through for maximum return on the project.

In preparing a contracting strategy, we need to think about the entire life cycle of the contract, as well as longer-term relationships with our client and our suppliers. Depending on whether cost, schedule or performance is the governing factor, we need to consider the end state of the project and how our client will respond to that end state. If performance is the driving issue, then, as a buyer, we should expect that factors such as our relationship with our customer are important. The performance of the project in the end needs to be well understood from our client's perspective. We should have been able to identify the scope and quality items that are non-negotiable compared with those that can be lived without if absolutely necessary and those that are arguably more correctly placed on a wish list rather than in a contract specification. A holistic view of the project that the contract forms a part of helps in understanding the project drivers and thus helps in selecting the contracting strategy that will drive the most appropriate behaviors.

A contracting strategy should consider every part of the contracting process, from how we select a vendor to how this contract will be closed out.

Probably the most critical part of a successful contracting strategy lies in the relationships that are created for its duration. Each relationship is governed in no small part by the way in which the relationship begins. This in turn is significantly influenced by the way we set about selecting our contracting partner. Every time we select a contractor or supplier to support the delivery of a project, we effectively make them a part of the success or failure of the project. If we fail to acknowledge their value and contribution, we are likely to disenfranchise them. This will lead to behaviors that reflect this type of dysfunctional relationship.

The most critical part of a contracting strategy lies in the front end, when we select the way we are going to bring other companies into our project team.

Bid or negotiate? This is a critical question. Negotiation implies more carefully considered selection of a partner than bidding does. Why? Because the bid process leaves a lot to chance and to a certain extent, to the luck (or otherwise, if a mistake has been made) of the contractor who had the successful low bid. Also, a low bid implies little – or more likely, no – room to move later on in the project when the inevitable wrong thing happens. Selection and negotiation, on the other hand, imply that the buyer knows the business it is in and can make a decision regarding the choice of a contractor. When we make this fundamental decision to work with one or more contractors or suppliers toward entering into a contract, we then contact the companies to invite them to bid or negotiate. This is our first contact with them and it often sets the tone for the rest of the contract. A modifier would be any past business relationship with that vendor. Established contacts have a big advantage and a big disadvantage. The advantage is that we know the people and the organization we are dealing with. The disadvantage is that we know the people and organization we are dealing with! Either way, this first contact influences what follows. If the first contact is positive and friendly, the outcome will likely be better than if that first contact is condescending, cold or even aggressive or defensive.

The first contract-based contact with a vendor is when it is invited to bid. This is the first – and key – contact that the contractual relationship is based on.

As the person who has to select – by whatever means – the supplier or contractor we will be dealing with, we need to decide on the process that will be used to make that choice and start on the path to a contract. Sometimes enterprise policies will preclude or limit the ways we have of entering into a business relationship with another party. Often the reasons for this are based on a combination of bad past experiences and the need to be auditable and conventional. All too often the choice of how we select a contractor has more to do with history and appearances than with what is suitable and appropriate for a particular project or situation. If the choice between bidding and negotiation is open, then such a choice should be made based on ultimate value.

Most practitioners know that buying based on the lowest initial cost does not necessarily get the best overall deal. Considering the cost of managing the contract, the potential for claims and disputes, and the extras and other factors in the evaluation of the bids will help provide a better comparison. That's the good news. Unfortunately, the process of making this evaluation is based on history with the contractor and is generally subjective, as well as a reflection of the relationship between people involved in the past. In other words, the process is not one that is easy – or even possible in some instances – to defend.

The choice of whether to bid or negotiate may be out of the control of the contract manager if it is dictated by policy. If it is within the span of control of those who set the contracting strategy, the decision should be based on value to the buyer.

If buying a commodity, price will be the primary factor to consider. If, however, the purchase includes a service, sustainability or other factor, then consider these factors when selecting the vendor. Services are NOT all the same, even if the qualifications of a number of suppliers of goods or services meet our criteria. The quality of service we get is largely dependent on the quality of the people who are involved. Quality in this case has more to do with the ability of the vendor's personnel to understand and respond to the issues that the buyer has. And this response is dependent on the buyer's representatives' willingness to share this information.

Buying a commodity is probably the simplest type of purchase. The product needs to meet minimum specifications. It needs to be competitively priced and it needs to be delivered at the most appropriate time. Selecting a vendor with these criteria is straightforward, all else being equal.

In some instances, the choice of a supplier or a contractor is relatively straightforward.

The type and scope of a contract should also be considered in determining the selection process for a vendor of goods or services. The more complete or complex the scope of a proposed contract (such as the inclusion of financing or operations), the more important the qualifications of the potential vendor will be. Other considerations include the value of any warranties, post-delivery service, supply of parts and user support.

Selection of a particular type of contract affects the choice of contractor selection process. Clearly, a stipulated price contract based on a complete scope and specification for what is required lends itself to competitive tendering. As soon as we deviate from this, though, we need to question the price competitive approach. If for no other reason, consider the ability of the buyer to fairly and objectively assess the effect of this shift away from unequivocal clarity in requirements on the bids received. If the scope is unclear, what are the chances that the successful low bid is from the contractor who will give us the least for our money? And having contracted with that organization, can we expect to pay more for what we still need once we have received all that the contractor intended to give us? And what is the additional administrative cost of getting this contract to work? If the specifications are vague or ambiguous, then a similar issue exists from the perspective of quality.

It gets worse. Consider now the need to include financing. This really changes the nature of the relationship as the vendor is now also, in a way, our banker, at least for the purposes of getting this work done. The traditional master-servant relationship will likely not work in this situation. The situation is much closer to a partner model. The same holds true for the deal in which the buyer is looking for the services of an operator for a built facility as part of the contract to build it.

The type of contract we select will affect a number of factors, including financing, payment methods, ability to influence the work and what we can change after the contract is signed.

Before we select a contract type and decide on the process for awarding it, we need to consider the extent to which we want to manage the work that is being undertaken. Consider a large project, such as a new theme and entertainment park. (I pick this as an example because, by virtue of its size, it serves to illustrate a few points that are easy to see. The principles apply to much smaller projects too.) This is a seven-year project to design and build this theme park. It will cost about $3 billion. It will involve design and construction of some high-tech rides, as well as special construction for unusual buildings. All the usual infrastructure features are present: water, sewer, roads, landscaping and lighting, power distribution and other regular construction. There will also be a need to include food and other concessions. Management of the facility, marketing and much more makes up the entire enterprise. The sponsor organization for this project may have a number of in-house skills that it will deploy in creating this large project. At the other end of the spectrum, it may have none. In the first situation, the sponsor organization will likely package and outsource the work it is not doing and will manage individual components quite easily.

A sponsor who does not have any of the expertise needed to complete the facility using in-house resources will need to outsource everything. The simplest way of doing this is to bundle everything, including operating the facility when it is built, into one contract. There will be remarkably few companies that can handle such a project, so the work may be negotiated with one of them. And the cost of this package would probably be higher and the risks probably unacceptable to the Sponsor or its investors. Alternatively, if it is offered for bids, consortia may put together teams to respond to this opportunity. Now we have a potentially complex contract between the sponsor and the prime vendor. We likely have a set of even more complex, possibly interlinked arrangements between the primary contractor, its partners and their supply chains.

If the buyer does not have the expertise to assemble the contractors, suppliers and others required to deliver a project, then it generally is faster and often cheaper in the long run to ask people who are in the business to assemble the team that can meet the requirements of the customer. The contractor will need to have the necessary expertise and connections to achieve success and will spend less time in recruiting and building the team, working out how to do the work and then delivering the product.

The scope of the contract often affects the amount of integration, coordination and subcontracting that the prime contractor needs to undertake. Larger-scope contracts generally save the buyer time and may save money.

One of the challenges faced by buyers is the ability to compare bids in a structured and auditable way. This has, over the years, led to what may be arguably one of the stranger practices in the buying process. Many owners like to have contractors bid on identical packages and either discourage or actively prohibit any creativity (i.e., any change from the original specifications) so that they can compare apples to apples. The cost of making the life of the buyer easier in this way is huge, as the opportunity to take advantage of the vendor's experience and expertise is effectively lost. Contractors and suppliers often do work for their customer's competition, so they probably repeat similar projects on a regular basis.

The greater the opportunity for a contractor to respond creatively to a request for a proposal or a tender, the more likely that the buyer will see innovation and opportunities for cost and schedule savings in the resulting proposal.

The trade off that the buyer needs to consider is between making the choice of one contractor over another easy versus trying to get the best value for the money being spent. In all too many organizations, the easiest and safest choice is to make the selection of a contractor as simple as possible. Value for money thus becomes a secondary consideration, as do the opportunities for the prospective vendor to be creative in offering its goods and services.

Preparing a request for pricing or proposals for vendors to respond to in such a way that allows for a degree of creativity is often a challenge because of the integration of components for which this one vendor may be contributing only one part.

As the opportunity to respond in a flexible way increases, so the ability of the buyer to compare proposals becomes more difficult.

Other factors that need to be considered in the selection of a contractor include elements such as risk, payment and the potential added value of intellectual property. Fortunately, many of the considerations we need to add to a contractor evaluation can be reduced to monetary equivalents. Table 4-1 lists some of the more common items that are included in a selection process.

Evaluation Criterion	Description	Monetary Equivalence
Base price	This is the base price for the scope and specification set out in the bid documents.	Base price may be adjusted to normalize against other bids if not exactly compliant with bid documents.
Potential to claim	This is a subjective assessment of the contractor's track record and the bid in particular. Experienced contracts administrators will look for limitations and other conditions that may be stated or implied in the bid.	This may be assessed as a percentage surcharge on the bid amount to reflect likely claims and the cost of managing this contractor over and above others.
Contractor experience	The direct and relevant experience of the contractor in addressing the issues in delivery of this project is considered. This is different from track record (see below).	No direct assessment
Alternatives to the bid	The contractor may propose a different solution to that identified in the bid document.	This is the savings (or additional cost) of the proposed solution compared with the original: Consider capital cost and the net present value of reduced operating and maintenance costs.
Track record on similar projects	This is a subjective evaluation of the contractor's capability to deliver what is required by the contract, based on past performance and client references.	No direct measure (Cont.)

Evaluation Criterion	Description	Monetary Equivalence
Proposed contractor team	This includes the resumés of the key personnel that the contractor proposes in its team for this contract.	No direct measure
Alternative components	Different equipment or materials may be proposed by the vendor because the vendor can get a better price for equivalent components, can get them delivered faster, has people trained in the installation of the alternatives, but not the originally specified components, etc.	This includes a similar assessment to that for alternative proposals.
Schedule compliance or acceleration	Particularly if the schedule is critical to the buyer, accelerated delivery of the contracted goods and services may have value to the buyer.	This is the cost of acceleration less the value to the buyer of earlier completion (e.g., revenues, reduced costs).

Table 4-1. Other considerations in selection of a vendor.

Choosing a contractor requires consideration of more than just price. Many, but not all, of these factors can be assessed in terms of a monetary equivalent. This assessment tends to rely on subjective opinion.

Particularly when the buyer plans to select a vendor based on a range of criteria, of which price may be just one, the rules for selection should be made clear to all bidders. There are two reasons for this. First, the bidders can respond more precisely to the issues of concern to the buyer. Second, the process is fairer to the bidders, who, after all, are spending time and money trying to meet the criteria to be successful in earning the deal. If the rules are not declared, the selection process is almost inevitably going to look unfair to those who are unsuccessful. In Canada, since the Ron Engineering Case and its fallout, it is a legal requirement to declare the basis on which a contract is to be awarded in the bid documentation.

To establish a fair approach to contractor selection, the whole process for tendering needs to be managed so that equity in the process and in selection is achieved. It also needs to be seen to be achieved.

One approach to selection of a contractor for engineering, procurement and construction services on a petrochemical project showed an interesting and innovative approach to the challenge of picking a project partner (see A Partner of Choice).

A Partner of Choice

A large multinational petrochemical company wanted to develop a project. They were looking for a partner who could provide the engineering, procurement and construction services. The project was viable only if both the capital cost and the cost of production of their product could meet the cost targets that allowed them to meet their internal rate of return objectives. From a short list of six potential contractors, the owner organization selected two who looked like they offered the best - possibly the most experienced – teams for the next step. This next step was to retain the two companies for a fee to produce concepts that may best address the business and technical issues involved and to present their approach to how this may be implemented. By paying the contractors for their ideas, the client organization obtained the rights to the intellectual capital in the proposed solutions. It then picked the preferred team, assessed during the process and based on the outcomes, and worked with the successful partner toward achieving the desired outcomes.

(Cont.)

> ## A Partner of Choice (cont.)
>
> In the end, commodity prices took off and the business focus moved from cost-driven to time-driven. The project was built with a stronger focus on speed to market than on the original cost-driven objectives. But the team to facilitate the delivery was already in place as a result of the initial work.

Contractors are being selected based on an increasingly broad set of criteria. The initial price is less significant today than are other considerations, such as speed of response, creativity and problem-solving skills, innovation, risk management skills, quality and so on. The challenge for buyers lies in devising effective ways of determining the capabilities of contractors or suppliers to perform under these different headings.

Some interesting innovations have been used in the selection of contractors. The nature and focus of competitiveness is changing; creativity, speed, risk and other factors are of increasing importance.

Given the growing interest in selecting vendors based on broader factors that affect the ultimate cost of a project rather than the initial price, the time that needs to be allowed for the selection of a vendor needs to reflect this. One step in the selection process is the response of the vendor to a call for proposal. This time has traditionally been set by the buyer and is usually based on a period that the buyer has allowed in the past. There is little or no connection between this time and market conditions, the nature of the request or other factors that will affect the ability of the prospective vendors to respond adequately.

Traditional times for the bid period sometimes bear little relationship to the time really needed to prepare a full and truly competitive bid.

Most buyers have been working on a project for some time before they invite suppliers of goods and services into the picture. They have had an opportunity to develop and become familiar with the nature of what they are building and how it may be brought into being, the challenges that this may create and how the final product of their efforts will be operated or used. Invariably this time frame is considerably longer than the time that a vendor has to get up to speed before submitting a bid or proposal. On a construction project, it is not unusual for the owner to have spent months, even years, developing an idea into a feasible project before asking one or more consultants (engineers or architects) to propose services for the detailed design. This team – owner and designer – may work on the project for more months, even years, before preparing the bid documents and inviting a contractor to bid. Even in a design/build, turnkey, or BOOT (Build, Own, Operate, and Transfer) contracting situation, the end owner will likely have spent a significant amount of time putting the project together and understanding many of its complexities before the vendors are involved. Compared with this, the time allowed to a contractor to understand the project and to prepare a proposal for their services is generally quite short. All too often we can measure this time in the number of weeks we can count using the fingers of one hand!

The time allowed for contractors to become familiar with a project and to prepare an effective bid is relatively short, especially compared with the time the buyer had in preparing to invite these bids.

In addition to understanding that time for a contractor to respond is short, we need to understand one more thing about times for preparation of bids or proposals. Unless the bidders had significant advance warning of an impending opportunity and had time to plan a response, there are practical constraints on the bidder. First, the bidder may have other proposals in the works that will tie up some of the critical resources needed to respond to this opportunity. Second, if there is inadequate lead time, the contractor may need time to react and mobilize a team to respond to the request for a bid or proposal or may not be able to bid at all. In the end, the bid period is effectively reduced by the time needed to react and the time needed to free up key resources.

The time allowed for contractors often exceeds the effective time available for the contractor to respond to an invitation to bid.

In the time available, the contractor needs to truly understand the intent of the buyer, as set out in the contract documents. This is important to the buyer – who does not want to select a contractor who is successful primarily because it misunderstood what it was bidding on. Equally, a contractor is unlikely to want to be tied to a contract based on a fundamental mistake. What follows is a discussion on this issue and how it may affect selection of the contractor. It is in obtaining this understanding that the need of common language and consistency becomes really important. It is in this stage of the development of a contractual relationship that there is significant room for planting seeds of future problems. One area that recurs on many projects is an understanding of the true performance expectations of the Owner. Performance is a balance of quality and scope. A reduction in the functionality of the end product may be a quality drop for the buyer but may be a scope issue for the vendor.

Physical elements such as wiring are more easily spotted than ones that have a less clear physical form until the intent of the description is understood by all – and is understood in the same way.

There is plenty of opportunity to create ambiguity (see Documenting the Documentation). We work with people who view the world differently or who make assumptions, ones who are busy and others who rely on a system that does not always exist, for validation or quality assurance. The result is confusion. The sad part is that we can afford all the time and money required to fix the outcome but we rarely have the time, money or patience to get it right the first time (usually the latter is significantly faster and cheaper).

The scope of work to be undertaken on a project is often unclear or ambiguous. This may lead to selection of the contractor who made the biggest mistake.

Clarity of scope in a contract is just one aspect of managing stakeholder expectations. Others include quality, safety, level of service, flexibility in dealing with inevitable issues as they arise, and more. If expectations are out of line with reality, friction will lead to dispute all too easily. Many of these aspects cannot realistically be articulated in the normal contract documentation.

A big part of a successful project is effective management of expectations. This is a two-way street!

Documenting the Documentation

Many projects today include sophisticated technology. Often the technology interfaces with other components and needs to be modified to suit the situation in which it is being deployed. This normally requires documentation so that users and operators, as well as those who maintain or upgrade the systems, will be able to do their work effectively after the product is delivered and commissioned. To address this, the specifications for the system will require the vendor to provide adequate documentation. What is this?

From a vendor perspective it may be enough to help the customer install the system and use it after some training. Does it include a help system? What about access to commented source code? How detailed do the user instructions have to be? Are there separate documents needed for maintenance? Or for people involved in subsequent upgrades? How specific does the documentation have to be to the application? Is standard system documentation good enough? What is "standard system documentation" anyhow? Normally these issues are resolved only after the work is done and someone is not happy with the result.

If we buy from a vendor who has different expectations from ours, we are probably buying trouble.

A previous bad experience in not having had expectations met often makes a buyer nervous. One response is to look for additional safety nets. One of the more popular forms of safety net is to require the contractor to be bonded. There are different types of bonds that are available.

If a buyer is nervous, there is a greater likelihood of bonds being required.

A bid bond provides a degree of protection against a contractor defaulting on its bid – usually because it made a mistake and ended up as the low bidder – and refusing to enter into a contract for the work with the owner. A bid bond normally pays the owner the difference in the price between the lowest compliant bidder (who has refused to enter into a contract) and the next bidder's price. The principle is simple: the surety that issued the bond covers the cost of going to the next bidder. The theory is probably sweeter than the reality, as we will discuss shortly.

A bid bond protects the owner from failure of a successful bidder to enter into a contract for the amount offered in its bid.

Concerns over performance can be addressed through a requirement that the contractor carries a performance bond. Often for at least 50% and sometimes for up to 100% of the value of the contract, a performance bond is issued by a surety to guarantee the contractor's performance under the terms of a contract. If the contractor fails to perform, then the surety that issued the bond will undertake to complete the contractor's obligations under the terms of the contract and to the limits set out in the bond agreement. As with all bonds, the process may take time.

A performance bond protects the buyer from failure of a contractor to perform its obligations under the terms of the contract.

The last type of bond is one that is used to protect the owner from having to pay the contractor's labor and materials suppliers in the event that the contractor fails financially during the course of the contract. In some jurisdictions, suppliers are protected from not being paid by the beneficiary of their work. Usually, in the United States and Canada, this is in the form of state or provincial legislation. This legislation, normally in the form of a builder's lien act or a mechanic's lien act, obliges the beneficiary of an improvement to have some financial obligation to pay for labor and materials expended in this improvement in the event that the prime contractor fails to make these payments. A labor and materials bond provides a degree of protection to the buyer in the event that this sort of situation and the resulting obligation occur.

A labor and materials bond protects the owner from the fallout and associated additional payments that may result from bankruptcy or other financial failure of the contractor.

If we consider the relationships between a contractor, its surety (bonding company) and the buyer, it is easier to understand why bonds are, maybe, not the best solution in every situation. First, the surety's client is the contractor, not the buyer. It is the contractor that selects the surety, pays for the bond and has to provide its own guarantees to the surety. This is a stronger and more permanent relationship than the one between the surety and the buyer, which really only comes into play in the event of a claim. What is the likely preference of the surety? Pay the buyer in the event of a claim or work with its client not to have to make any settlement unless it really has to? If the contractor is clearly not performing or is clearly in trouble financially, payment may be made. Under many circumstances, though, it makes more sense not to sour the more permanent relationship. This is easily done, as most sureties require substantial commitments in the form of a large deductible, that are ultimately payable personally by one or more of the owners or senior executives of the contractor organization. These senior managers are put in a position in which their car and home are routinely on the line, together with other personal assets, in the event that a claim is made.

In the end, this all translates into a reluctance to move quickly or willingly to resolve a claim made under a bond unless the reasons for settling are quite compelling and unequivocal in the view of the surety company.

The theoretical value of a bond is probably much greater than the real value in the event that the bond is called.

4.2 The Bid Document

In Section 4.1, we looked at the process of selecting a contractor. Often we ask prospective contractors to submit a bid. A document of some sort is usually used as the basis on which we solicit such bids. What needs to go into such a document? Essentially, we need to include those elements that the supplier needs in order to make a reasonable offer to us as a buyer. When we look at the document from a supplier's perspective, we need to have clarity around what is needed. In the absence of such clarity, we need the freedom to meet with the buyer and to understand what he or she is looking for. In the absence of even this, we would need some freedom to be creative in responding. More specifically, we would need an opportunity to sell to the buyer, because what we are really trying to do in such a situation is to ensure a place at the table when the negotiations begin.

Put another way, what anyone trying to sell goods or services needs is to know what the buyer is looking for and how to best meet that need. If the bid documents do not provide clarity around the needs and any specific terms and conditions that apply, the responses from the bidders will be disappointing.

Included in a typical set of bid documents are the following main elements:

- Instructions to bidders (e.g., how to submit a bid, what will be considered)
- Commercial terms (e.g., payment method, specific expectations regarding insurance requirements, bonds, compliance with corporate or regulatory requirements)
- General conditions (e.g., contract administration process)

- Detailed or performance specifications (e.g., description of objectives, technical specifications, drawings, concepts, primary quantities, technical limitations and issues)
- Schedule and other requirements and preferences (e.g., deadlines, including intermediate ones, coordination with other activities or suppliers, process expectations)
- Project-specific information (e.g., soil conditions, as-built drawings, available facilities, site restrictions, limitations for access or working hours, interfaces with existing technologies).

The key ingredients of an invitation to contractors and suppliers to make an offer should be present to minimize risk of a future problem with the contract. Miss critical information in a bid package and the probability of future problems increases.

Instructions to bidders should include any specific approach that the buyer will use in assessing which of several bids will be selected. In the absence of these declared rules, several things happen. First, the bids may miss the mark in terms of what is important to the buyer. Second, there will be a sense of unfairness in the selection process by the unsuccessful bidders. Third, the unsuccessful bidders will think twice before spending effort and money to respond in the future.

Instructions to bidders provide the basic information necessary for the bidders to understand the rules for submitting a bid and the rules used for selecting the successful bidder. These rules are particularly important if the selection is based on anything other than just the price.

Apart from the confusion caused with any bidders, the absence of clarity in the instructions is often an indication of the absence of clarity in what the buyer really wants. This makes picking one vendor over another much more difficult.

If the instructions to bidders are unclear, the clarification and evaluation process becomes difficult and time-consuming.

Whatever we ask the bidders to do, we need to remember that we are asking them to do something – bid – without paying them. Only one bidder will be successful (and therefore be rewarded or compensated); the others will not. Some examples of possibly unreasonable behavior include asking too many bidders, expecting a response in an unreasonably short time and demanding priority over other clients that the bidders may have. The end result of such requests is that some or all of the bids will reflect a similar sense of unreasonableness, either limiting the number of truly competitive bids or increasing the cost of the goods and services.

 Instructions to bidders need to be realistic.

The commercial terms set out all of the business aspects of the deal. Typically, the commercial terms include payment amounts, how to request payment, timing of payments, holdbacks, interest charges, agreed amounts of mark-ups and so on. Also included in this set of information are the expectations for delivery of goods and services, which may include what is covered and what is not covered under the terms of the contract. For example, if an item is delivered, does the price include shipping, insurance or delivery inspection?

 The commercial terms of a contract set out delivery expectations, payments and other business expectations of both parties.

It is normal for the commercial terms in a call for bids to be written by or for the buyer. Often these terms are non-negotiable in a bid situation. Although this has been the approach for a long time, it does create a master-servant relationship from the outset. As buyers, we prefer this type of relationship. As vendors, however, we prefer to be treated as partners rather than in subservient roles. This is the human condition at play. Part of the human condition is also the need to trust and a natural resistance to doing so. This creates a potential at the outset for subsequent confrontation if the relationships in a contract are not well managed.

Often the commercial terms are written by the buyer or the seller with little or no discussion with the other party. This take-it-or-leave-it approach reduces ownership and increases the potential for confrontation.

The General Conditions of a contract are also normally written by one party – usually the buyer or its representative. These conditions include the "rules" for management of the contract during the process of meeting the obligations of the parties. Typically, the general conditions of contract include the following:

- Documents and additional instructions
- Role of third parties (e.g. consultants)
- Delays
- Changes in the work
- Remedies for failure to perform
- Termination of the work
- Disputes
- Assignment and subcontracting of parts or all of the work to be performed
- Coordination with other contractors and suppliers
- Processes for payment
- Applicable taxes, laws and regulations
- Insurance and indemnification, warranties and bonds
- Other administrative and special needs of the buyer or seller.

This list is somewhat generic, as each document tends to be specific to an industry, sector, or even a company. There are several industries that have developed standard General Conditions of Contract for various forms of Agreement that include most of these components.

The General Conditions provide the administrative framework for the Contract. These conditions describe how the most common occurrences will be addressed.

What follows was covered earlier in a different context. It is repeated here for clarity.

Another aspect to consider when preparing the commercial terms of a contract is the role and involvement of the contractor in defining and delivering the specified goods and services. As with most options in contracting, there is a continuum to consider. At one end, the Buyer tells the Vendor exactly what it wants. Detailed and very specific technical specifications describe what the buyer wishes to acquire. At the other end of the spectrum, the buyer describes what it wants to achieve, defining the goods and services in terms of outcomes rather than the components that it wants to acquire. Whereas the first approach implies a solution – and the Buyer is responsible for its effectiveness – the second approach places responsibility for the effectiveness of the solution on the Vendor.

The most common approach is to develop detailed specifications first, then use the specifications as part of the bid documents to obtain competitive pricing for delivery of the goods and services required to meet those specifications. If a contractor then processes the work based on these detailed specifications, the Contractor should be expected to deliver to that standard. The counterpoint is that the Contractor should not be expected to deliver either more or less than what was specified. Whether the end product does what the Buyer expects or not is immaterial in this situation.

Detailed specifications provide the Contractor with precise instructions regarding what the Buyer wants. The contractor will only deliver what is specified.

A performance specification, on the other hand, is based on describing what we want to achieve, then asking contractors to propose solutions, which may include a guaranteed price for delivery of what they propose. This allows the contractor to devise solutions based on a defined problem. There are more degrees of freedom in such an approach. The opportunity for the contractor to be creative and offer innovative ideas now exists.

A performance specification describes the outcome and allows the contractor or supplier some latitude in how it is going to deliver the outcome.

To better understand the difference between detailed and performance specifications, consider building a summer cottage. The performance specification may include statements or concepts such as "low maintenance", "easy to live in" and a need to accommodate up to five adults in comfort. The detailed technical specifications may include specific solutions to problems that have been encountered in previous cottages. So the plumbing may be designed to drain easily to avoid burst pipes in winter, when the cottage is not in use. Because of experience in other situations, all subfloors may need to be screwed, not nailed, in place, because squeaky flooring makes the place hard to live in. The detailed specifications may specify three bedrooms, two with en-suite bathrooms.

Apart from the obvious difference that performance specifications are based on expected outcomes and detailed specifications present solutions, there is one other important difference. If we want to use detailed specifications, we need to have the experience in designing and living with the type of product we are buying in order to address the detailed specific solutions to the problems we are solving.

Thus, it makes sense to consider performance specifications in situations where the vendor either has the experience we lack directly or has helped others solve similar problems and is aware of the potential solutions.

Performance specifications make more sense when we are also buying experience and expertise. If we really do know what we want, a detailed specification will help achieve that objective.

One place where experienced owners have tried to be more creative is in the development of North Sea oil and gas fields. Through the Cost Reduction Initiative for the New Era (CRINE) in the United Kingdom (and its counterpart – NORSOK – in Norway), innovative contracting approaches have taken advantage of the operating knowledge of the owner organization and the design and construction expertise of the contractors and their suppliers. Owners in this arena were used to solving their own problems and telling the contractor exactly what to build. Contractors were used to doing what they were told to do. A conventional approach to finding and developing oil and gas deposits made many of the prospects in the North Sea non-viable. So the problem was turned around. Contractors were asked to propose solutions that fit a viable budget. If they could deliver what was needed for that price, then the project would go ahead. In the end, many creative solutions came out of this approach. Some are listed below.

- Re-use installations on different sites, designing them to be more mobile (e.g., production ships instead of production platforms)
- Lease equipment rather than buy it (using off-balance sheet financing and allowing the competition to use the same equipment on their development later, reducing the capital cost and managing the operating cost)
- Remove some of the engineering redundancy in designs where such redundancy had not historically been used on other projects.

Some of these solutions challenged assumptions and practices that were previously held sacred.

Experience in the North Sea has highlighted some of the advantages of performance specifications, especially if the specifications themselves are open to honest challenge.

British Petroleum wrote about some of their experiences in a book by Terry Knot entitled, "No Business as Usual" (1996).

Schedule and other requirements and preferences also need to be considered. If something is important to the buyer, it needs to be articulated in the invitation to bid. In its absence, a contractor striving to be the lowest bidder will not assume we want it – it will only add to the price, which means they will not get the work. In other words, the lowest bidder probably offers us the least. That equates to exactly what we asked for. Certainly, it is unlikely that they would have included something we did not ask for (even if we should have) – especially if we are going to select the Contractor on price or if price is a significant consideration in the selection process. This in itself creates a challenge if we need to compare bids from different contractors for anything that is more than just a commodity.

It is important to think carefully about what we are buying. Any expectations that are not articulated in the bid document will likely not be delivered by the seller – especially if the contract is to be awarded based on lowest cost.

4.3 Bid Comparisons

In Sections 4.1 and 4.2, we discussed the process for selecting a vendor and the documents we need to issue to be effective in this selection process. One would think, therefore, that comparing bids would be an easy thing. This is theoretically true if the bids themselves were identical and the capabilities and reliability of the potential suppliers of goods or services were equally consistent. They are not. So a number of ways of making comparisons that lead to selection of a contractor have been devised over time.

Setting the rules for bid comparison is a critical step in the process. Basically, if the rules are unknown, there is no basis for making a comparison. Equally, if the rules are unknown, the contractor cannot respond as well to meet the criteria and expectations. If we are regularly in the business of buying goods and services, then we are more likely to get truly competitive and responsive bids from contractors that trust our ability to assess them fairly and who have confidence in the process we use. Anything in the selection process that is not transparent becomes a mystery, which leads to the impression – justified or not – that the buyer has a hidden agenda. At worst, this could imply or mean that the process is unfair or predetermined.

It is important to make sure the rules for comparing bids are clear. Also, the bidders probably should know what these rules are before they submit their bids.

If we compare prices for two identical items, it makes sense to select the cheapest. As soon as the items stop being truly identical, or if service, timely delivery, warranties and other factors are important, then a direct price comparison is arguably no longer valid.

If the only consideration is the lowest initial price, then comparing bids is simple.

Wise buyers consider factors other than just the price when selecting a supplier of goods or services that are anything other than just a commodity.

Often a fair evaluation is difficult to achieve when comparing contractor bids. One of the biggest reasons for this is that we need to deal with what some view as intangibles. One such example is the ability of the contractor to successfully complete work by a specified date. If delays beyond that date are likely to cost the buyer a lot of money (lost opportunities, lost production, missed revenues, interest charges, etc.), then a reasonable assessment of the potential for the delay and its effect on the overall project needs to be considered. Dealing with intangibles such as this requires a way of assessing the variables involved. This may also require identifying what those variables are and what their likely effect on the project may be.

For example, if one vendor offers a more competitive price but cannot (or is unlikely to) deliver on time, whereas another is a bit more expensive but can safely be assumed to deliver on time, do we still pick the cheapest? We need to understand more than just price in this process. We need – in this instance – to also understand the risks and their likely effect on the project overall.

Finally, we need an auditable way of describing these situations in the selection process or we need to be extremely experienced and confident to justify anything that is other than the "obvious" choice. This reinforces the need to have the selection criteria established at the outset, so that there is no appearance of tampering with the results.

Assessing the value of intangibles such as service and potential for problems usually requires a significant amount of experience and courage.

We have just looked at a situation in which we do not award a contract to the lowest bidder. The most obvious reason for doing this is that the bidder has submitted a non-compliant bid and therefore is disqualified. If this is not the case, then the assessment of how the successful contractor or supplier is selected needs to be clear and tamper-proof if the people involved in the decision are to be protected.

It is difficult to award a contract to any bidder other than the lowest one unless the rules are established ahead of time.

The need for a "safe" contractor selection process may lead to an unwelcome result. Look again at the selection criteria in Table 4-1. Those items that have no monetary equivalent are the ones that make the selection process difficult.

The need to be auditable and appear fair in selection of a supplier after some form of bid process, can lead to award of a contract to the wrong party.

4.4 Bid Process

The process of bidding is better understood from a buyer's perspective. It is the *bidder's perspective* that throws some useful additional light on what really goes on. There are a number of steps that a typical bidder will go through when making a decision to offer a bid on a piece of work. These are not always followed, nor are they taken in sequence. This list of steps may overstate what some companies do and clearly is a simplification of what large and sophisticated companies follow in preparing bids for significant projects.

The first issue that is likely to be addressed is whether the opportunity is one that the company should be entertaining in the first place. Is this of strategic importance? Is it in our line of business? How will it affect our other work? Is this a real and worthwhile opportunity for us?

Do we want the work? (Vendor's question.)

If we get past this first question, then it is worth assessing the chances of success. There is a cost (both a direct one and an opportunity cost) associated with preparing and submitting a bid. If, for example, there are several opportunities on the table, which ones are likely to bring the greatest return? When are we the strongest contenders? If there is a bonding requirement, we may have limited capacity and need to pick which projects we bid on, so as not to exceed this limit. Or we may need to negotiate an extended limit with our surety (bond provider). Basically, if the effort is too great to justify or if the field of competitors is too strong, we may decide not to bid. The strength of the field is not measured just in terms of capability, but also in terms of how hungry (desperate?) we think they may be. Do we really want to bid against one or more companies that are known to be desperate for work and will go to extremes to get it? Perhaps a worse competitor is one who lacks experience and knowledge of the project in question and is likely to make mistakes. They are likely to forget or miss something with the inevitable result that they will be awarded the work. The only mitigating factor would be the buyer's ability to recognize, and deal with, such an incomplete bid. This is partially identifiable through the selection process, should the buyer have stated what this is.

 Who is the competition? (Vendor's question.)

Use of a fair and manageable selection process is certainly one indicator that will help identify a reasonable client. Other factors to consider include the ability and willingness of the client to pay for the work that has been done, the fairness of the buyer's contract administrator, the level of bureaucracy and whether that contract administrator knows how to work within its limits. Once we have made a preliminary assessment using these considerations, then we can assess the nature of the bid documents (complete, unambiguous, fair) and the way in which the buyer deals with the bidders. Consider the following.

- Are we on the inside track?
- Will we be treated at least as well as other bidders?
- How well do we know the people involved?
- Do we suspect that the documents and process are geared toward selection of a predetermined contractor? (Good if it is us, bad in all other cases!)

Overall, the assessment we make here is one that helps us determine if we really are prepared to work with this prospective buyer and what we will have to do to keep the client happy. Only this latter part is important, as it may affect the cost of working with the client (staff time, meetings, legal costs, entertainment and more).

 Is this a client we want to work for? (Vendor's question.)

Now we have an opportunity we are interested in and a client we want to work for. The next question is whether it is worth our while working for this client on this project? Are there better opportunities elsewhere? Have we generally been profitable when we worked for this client in the past? Is this a relationship we want to preserve? How high are the risks on this project? Does the proposed work line up with where we are going as a business? The bottom line, really, is do we really want this opportunity? And if so, what do we need to make out of this (not just in profit but in future work, a new experience, a referenceable project, etc...)?

 What is the opportunity worth to us? (Vendor's question.)

We have already asked the question about whether we will get paid by the client or not. The first time we asked, we were probably thinking about whether the client could afford the project and had the money to cover the cost. Now add another dimension. How likely is the project to change? And if it does, or if for some other reason costs increase after the Contract has been signed, will these additional costs be recovered?

 Will we get paid? (Vendor's question.)

Part of the answer to this precious question lies in how easy it will be to work with the client after the contract has been signed. In addition to change orders, the nature of the relationship needs to be considered, as this will affect whom we assign to the project to manage delivery of our goods and services. The extent to which we feel the need to protect ourselves will, in part, be dependent on whether we have a good working relationship or whether we are in a confrontational master-servant relationship.

How easy is the customer to deal with? (Vendor's question.)

Just as we looked at the level of desperation or determination of the competition to win this work, so we must examine our own position. To what extent do we need this project? What are the implications of not getting the work? What are we prepared to risk to win the contract?

With that accuracy, you MUST be the person who prices our change orders.

How desperate are we for this sale? (Vendor's question.)

To avoid being like the dog that caught the car and did not know what to do next, we need to be sure we have the capacity, capability and resources to complete the project successfully in the event that we earn the right to do the work.

 Can we do the work? (Vendor's question.)

The person who will manage the project if our bid is successful should probably be a part of the decision process and be involved in closing the bid. There are several reasons for this. First, we need to know who this person is if we want to invite him or her to participate.

 Who will manage and deliver the project? (Vendor's question.)

The bidder on an opportunity is usually just one link in the supply chain. It is common for contractors to rely on specialist subcontractors and suppliers. This means that, to be successful in their bid, they may need to involve at least some of the key members of this supply chain in the preparation of their own bid. The bidder then becomes a buyer. Subcontractors and suppliers will likely go through a screening process similar to the one that the bidder went through in deciding whether or not to submit a proposal.

If we need to rely on subcontractors and suppliers, we will need to consider their ability and willingness to respond to a request for proposals or a call for bids. (Vendor's issue.)

The relationship between the contractor and its subcontractors and suppliers is probably one that has a significantly longer life span, covering several projects over a number of years, than the relationship between the contractor and the buyer, which may be new or for the duration of only one project. Contractors rely on good relationships with suppliers and subcontractors for sharp pricing and flexibility in working around the inevitable glitches that occur on a project. This relationship and ability to work effectively are key to maintaining a real competitive edge. In the absence of any established relationship, this deal is just another in a series that involve normal competitive pricing. The difference lies in marginal lower cost associated with the confidence that we will be treated fairly, and that there will be an exchange of ideas and some creative solutions that come out of this, including use of alternative materials, better sequencing of work and more effective timing of critical events.

Subcontractors and suppliers consider their past relationships with a contractor when making an offer for supply of goods or services. (Everyone's concern.)

The buying process in many organizations is managed in such a way as to minimize the risks to the organization itself. This often translates to a fully auditable process in which the ability to make decisions at the operating level is severely limited by the need to be seen to do the right thing. All too often, one of the "right things" is to pick the supplier who offers the lowest initial price. This was generally the practice in one of the companies I ran until we hit a problem more than once. This problem was that the low bidders for two specialist services that are commonly found on the type of project we did were suing each other on another project. The acrimony and negative relationship spilled over onto our project. This led to delays and what appeared to be deliberate sabotaging of each other's work. It cost us a lot to sort out the mess, adjust sequences of work so that these companies saw as little as possible of each other on our project, and ensure that the hand over of work by one company to the other was kept reasonable and fair. The cost was significant. We felt it exceeded the difference between their price and the next bidder. Our policy changed. Instead of just lowest price, we looked at existing relationships between our contractors and their ability to cooperate.

Some subcontractors who do not get along may end up working together on a project for a prime contractor. It pays to understand existing relationships between other corporate members of the team. (Mainly a vendor issue, but affects the buyer in the end.)

Some of the trades in construction can cause anxiety or problems to the unwary or inexperienced contractor (see Mysterious Metals). The lesson is the same one that buyers struggle with: The lowest bid is really not always the cheapest in the long run.

Mysterious Metals

In the business of assembling a price for a general construction contract, it is common to solicit prices from specialist subcontractors and suppliers. One of these specialists is a company that does all of miscellaneous metals work. This is a collection of metals that typically need fabrication and installation, but that are not provided by the structural steel contractor or by other subcontractors in the normal course of their work. If this sounds like a vague description, it is with good reason.

Included in this group of materials is a long and eclectic list of items that includes special handrails, canopies, decorative metals, special cladding for specific parts of a building, designated metal housing for components and much more. Similar to the bidding of other subcontractors to a general contractor, the prices will come in at the last minute to minimize the chance of the general contractor shopping around for better pricing. Typically, general contractors pick the lowest bid. Some time after the bid has been closed, the fax starts humming again. Many of the messages come from the miscellaneous metals supplier. Their faxes list all the stuff they excluded from their bids. Unfortunately, this is too late for the contractor who has already submitted a bid to respond by adding the additional cost of the excluded items to its bid.

(Cont.)

Mysterious Metals (cont.)

It is in just this kind of situation that the general contractor's estimator earns his or her huge salary. The best estimator will have clarified what is to be priced and will have stated that any price will be deemed to have included all of those items. Problem solved. If a low and unsolicited price arrives, the estimator responsible for closing the bid will get clarification from that subcontractor or, if there is not enough time, will assess what may be missing from the bid and adjust the price to suit. Either way, and despite the best efforts of a well-organized estimator, this trade often leaves a small cloud of uncertainty around what exactly its price really covers. Which is why this particular trade contractor's scope is often referred to as mysterious (not miscellaneous) metals.

The best overall deal from a subcontractor, like that from a prime contractor, is not necessarily with the lowest initial bidder. (Everyone should be aware that the same principles apply for both the buyer and seller.)

Because of the problems created by post-bid clarifications (see Mysterious Metals) and other factors that contribute to effective price changes, contractors need considerable skill in assessing the real value and price of a bid.

Normally we would worry about a bid that was particularly low. First, such a bid is tempting to accept, as it lowers our overall contract price. Conversely, if we do not accept it, or if we adjust it because we believe it to be too low, we may price ourselves out of the market. Also, if our competition does not pick up on the exceptionally low price and carries it in their bid, they gain an advantage at the bid stage (i.e., they get the work) even if they also get a disadvantage executing the work (i.e., they lose money). Some of the other factors that need to be considered include the following:

- Has this subcontractor or supplier really understood the scope of the work?
- Does the price reflect innovation or a better approach than others may have?
- How reliable has this vendor been in the past?
- What is our relationship with this vendor (i.e., are they helping us)?
- How well does the vendor work with those who do work before or after this project? (The interfaces may be a problem and an opportunity to present a claim if not managed well.)
- How well is the work (scope and quality) defined in the bid documents?
- Are there special needs that this contractor has missed (e.g., several site visits being required, risks that need to be assumed or unusual conditions on site or in the contract)?

If time permits, it usually pays to get the low bidder to clarify and confirm a price. If this is a higher price, they need to either withdraw their bid to other competing contractors or adjust it appropriately. If they fail to do so, our bid will be disadvantaged.

Before closing a bid, the final price needs to be adjusted by the contractor to reflect risk, market conditions and other factors that may influence the price or that are likely to affect the performance of the required work. (Vendor's issue.)

General contractors also have to deal with numerous last-minute price collection issues, including:
- Last-minute changes to the bid documents coming from the designer
- Modified bids coming from subtrades
- Exceptionally low-priced alternative solutions that require coordination with other trades to fully incorporate, but there is insufficient amount of time to do so
- Low bids withdrawn or modified at the last minute by subtrades

- Qualified bids from suppliers that do not fully comply with the specifications but are likely to be carried by one of the competitors because the price is lower
- New information that invalidates assumptions made.

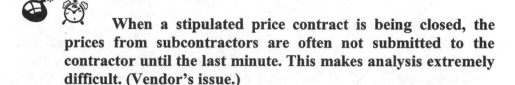

When a stipulated price contract is being closed, the prices from subcontractors are often not submitted to the contractor until the last minute. This makes analysis extremely difficult. (Vendor's issue.)

The bid depository system was established in North America to address some of these bid closing issues. In particular, the challenge was to help the mechanical and electrical subtrades, whose work often represents a very significant part of the overall contract. Also, the work they do is normally designed and specified by specialist engineering companies that are subcontracted to the prime consultant. The approach is simple in principle. These key trades (which can include other significant trades on large projects) submit bids to the general contractors through an intermediary. The closing date is typically 24 hours before the main contract closes. The bid depository (intermediary) holds copies of their bids in escrow, so the traditional "games" cannot be played, such as bid shopping, comparing bids by involving the competition or haggling over prices. The general contractor receives the bids from the bid depository. Subtrades have no obligation to submit the same price to each of the contractors.

One attempt at solving the last-minute closing of subcontractor bids has been the use of a bid depository system. (Vendor's issue, but often a Buyer's concern.)

If there is a question as to why I have not taken a buyer's view of the tendering process, then I have failed to make a point. The general contractor faces many of the problems a buyer faces in selecting its subtrades and suppliers. The key difference is that it often happens faster. Furthermore, the general contractor should be an expert and should have established relationships and some knowledge of most of the bidders it works with.

General contractors understand the business of buying as well as selling because of where they are in the construction supply chain.

4.5 Games People Play

Section 4.5 provides some of the insights that most of us gain over years of experience and exposure to the human condition in competitive situations. Similar to the way we need to do drug testing in competitive sports, so we need to do equivalent testing in competitive bidding for work. Jobs, retirement funds, homes and more are routinely at stake in the business of competitive tendering of work. Sadly, practices that are unethical, and in some cases downright illegal, do occur. In Section 4.5 we look at some of these practices, so we can more easily recognize them and address them.

There are, as may be expected in any competitive situation involving humans, a number of games that are played in the interests of survival.

I already mentioned price shopping in Section 4.3. Exactly what is this? It is the practice of taking one supplier's price and calling other suppliers, disclosing the price of the competition and asking for a better deal. This approach damages the reputation of the shopper in the eyes of the bidder. Usually it is self-policing, as any supplier who has been burned this way in the past will adjust its price (invariably upwards) for that buyer. That way they can be sure that their price is somewhat protected, as it is less likely to be the lowest. If they really want the work they can compare the shopped price of the low bidder with their real number and decide whether they want to do it. This practice, too, is unethical in my mind. But it is probably better than nothing, as the contractor who does the price shopping is now less likely to get the work, as most of their subtrade prices will be higher.

Unfortunately it is not that simple. An unscrupulous contractor will know what is going on and will adjust the trade prices it gets downwards, counting on negotiating a final price with its subtrades after it has won the work. This is one of the reasons that the bid depository was established in the first place.

Look out for contractors that change their subtrades on a frequent basis after they have been awarded the work. There are lots of reasonable arguments – often valid – for changing a subtrade later. These include the inability of that subtrade to really do the work, shortage of resources, schedule conflicts, quality concerns, union conflicts that were not known at the time of bid and more.

Price shopping is a practice that is rightly frowned upon by most professionals in the contracting business. (Everyone's issue.)

Another interesting situation is the one in which a contractor is asked to submit a bid on a project, but for some reason is not prepared to do so. At the same time, the company does not want to upset the client or its consultants. The solution is to submit a nominal bid. This is done by getting a "safe" higher price from a competitor then submitting that number as a bid. It requires almost no effort to put the bid together and provides a price that is in the right ballpark compared with at least one other bidder. This verges on price rigging, but with the intent of saving effort and trying not to upset a current or prospective future client by overtly saying we are not interested in their work.

A contractor can expect favorite subcontractors who are too busy to support the bid to take similar action. This clearly requires a significant level of trust between the competitors who are willing to exchange this type of information. If the nominal price is submitted with appropriate care, then the recipient will not even know it has happened.

Sometimes the anxiety of not submitting a bid to a past or prospective future client is addressed by submitting a nominal bid. Unless this is bungled, it will be hard for an owner to spot that this has happened.

The nominal price is arguably on the acceptable side of a very fine line in intent, if not in practice. On the other side is collusion. Collusion is deliberate rigging of a bid by the bidders such that the outcome is predetermined. The objective is somewhat sinister, as the people who are collaborating decide on who will win the bid, which will usually be at a margin significantly above market prices. Price rings are generally (and rightly) illegal in most civilized countries for obvious reasons.

Price rings and other forms of collusion do occur. We read about them in the papers and we hear rumors in industry. Clearly, this is illegal in most jurisdictions – for good reason.

Another potentially difficult situation is submission of an alternative bid. This is always an option for a contractor, albeit a risky one in certain situations. The opportunity lies in saving the client a significant amount of money with an innovative solution. If the bid process precludes submission of "non-compliant" bids, there is a risk that the contractor's effort is wasted, as its bid will be rejected out of hand. However, should the bid be attractive enough, it may tempt the owner to negotiate with the innovator or possibly to request new prices from all of the bidders based on the innovation. In Canada, because of legal precedent, this would prove to be difficult. In other jurisdictions, this may be legal, but the ethics are clearly questionable. Despite this, it has been known for owners to take advantage of the proprietary information that is submitted in such alternative bids. To prevent this, contractors have made offers based on a high-level concept that they disclose only in final contract negotiations. Even in this case the risk remains that the negotiations "fail" at the last minute, and the owner is then free to do what it pleases, unless it has signed a non-disclosure agreement.

It is interesting to note how, once we start to mistrust, we end up with many additional steps in the contracting process. This issue is brought up again in Chapter 7, in which we look at the role of trust in contracting.

Alternative bids are hard for some buyers to deal with because of their bidding process. (Buyer's issue.)

Alternative bids can offer innovations and creative solutions not originally considered by a buyer. (Buyer's issue.)

One of the challenges that exists in the submission of an innovative solution by a contractor is that the innovation itself may prove to be a cause of embarrassment to the consultant whom the owner retained to design the project. The obvious question is why did the consultant not think of this solution in the first place. This may lead to debunking of the notion presented by the contractor to save face. In some cases, the situation becomes difficult to disentangle, as both positions are arguably valid. One such example is the proposed use of precast concrete instead of the traditional poured-in-place concrete to build a waste-water processing tank. The proposed design had not been built before, but it did meet the building and design codes. The proposal would have saved a significant portion of the budget for the project, but it was rejected by the consultant and the owner because it was an untried idea – even though the contractor offered guarantees. This may in fact not be a good example. Had it been built, would it have worked? Now we will never know.

Really ingenious solutions offered by a contractor can be an embarrassment to a design consultant that came up with the original solution that the contractor is challenging. (Buyer's issue.)

Another form of variance to a standard contract bid is taking exception to specific bid requirements. There are choices that the contractor may make in this case. Clearly one option is not to bid on the work. Another is to submit a non-compliant bid. A third option is to offer a competitive price based on the absence of the term or condition with a significantly less competitive price should the condition remain.

Sometimes contractors are reluctant to comply exactly with the terms and conditions of a bid request. In this case, they may decline to bid, state the exceptions or adjust their price to reflect their concerns. (Buyer's issue. This makes it harder to understand the spread in prices from bidders.)

The example "So you think you got a deal?" in Section 2.3.4 shows the effect of adding a risk clause to a contract. The rationale behind adding the risk clause may well have been driven by a similar clause in the contract between the owner and the engineering, procurement, and construction management contractor. The ultimate buyer paid dearly for the apparent protection that the clause afforded. Thus the issue is not a simple one. Often buyers rely on legal advice that undoubtedly provides good protection. The cost of that protection may far outweigh the benefits, however. Assessing the cost and benefit is not easy and is something that the astute buyer should consider and get appropriate advice on.

 It pays for a buyer to really understand the implications of how they purchase goods and services.

4.6 Negotiations

After an initial offer from a supplier, it is not uncommon to go through a negotiation phase. Whether this is simply to gain clarification or whether it is to see if a better deal can be wrought from the process is sometimes a bit unclear.

Clarification of a bid is a reasonable and often sensible step before signing the contract. What such a process should do is facilitate a meeting of the minds between the two parties. Once the formality of the tender process is complete and before the contract is finalized, it makes sense to eliminate any potential for misunderstanding. If the owner or its representative has concerns, such as a low price or unit rate that may be an indicator of a misunderstanding, it is better to resolve or eliminate the underlying issue. Conversely, during the bid process, a number of potential misinterpretations of the intent of the owner or the designer may have arisen. Again, gaining clarification before entering into a contract normally pays, simply because the courts tend to find against the author of a document in the event that there is ambiguity.

A straightforward bid clarification should simply help reduce ambiguity and manage the expectations of the buyer and the seller.

Bid clarification meetings should be carefully minuted, as the decisions and points agreed on may affect the terms and conditions of the contract. The normal objectives of meetings of this nature is to clarify specific details in the bid documents and to ensure that both the buyer and the vendor understand and intend the same thing. Done well, this is an opportunity to make necessary adjustments to the documents for added clarity before signing the formal agreement.

It is worth remembering that a bid clarification meeting forms the basis on which work is done from this point onward under the contract that is being finalized.

Review of scope and other factors such as timing and expectations for safety, access to key resources, information and quality expectations are just some of the items that are commonly covered in such clarification meetings. If the project is very conventional and is being done in an area where both the buyer and vendor have worked before, then there is likely to be a shared understanding of what may be expected from both parties, as well as what local practices, standards and traditions may be. As the project enters new areas (technology, location, nature of work, unusual timing or sequencing etc.), then the probability of potential misunderstanding or misaligned expectations increases. Knowing this will help in planning for, and managing, any such meeting.

One good practice is to work with checklists based on the things that have gone wrong in the past. Over time, such checklists can only improve as we add new issues and learning from experience.

The more traditional the work or items being supplied are, the easier it is to avoid misunderstandings about what is expected. Without the underpinnings and implicit knowledge of such tradition, the opportunities for misunderstandings abound and need to be addressed.

Cost-reduction options may have been part of the bid that is now being clarified. These options and possibly some new ones may be presented and discussed at this meeting or some other pre-contract award meeting specifically arranged for this purpose. There may be several reasons for this meeting. The meeting may simply be to explore potential cost-saving solutions. It may also be because the lowest price offered by a bidder was still higher than the amount that the buyer was able to spend on the project. Here, again, there is opportunity to mess with the bid process, so some organizations "protect" themselves by negotiating only with the successful low bidder. Others will interview more than one bidder and, after adjusting prices to reflect a real balance, will then select the true low bidder.

In addition to these options, there is a full range of marginal practices that include price shopping and hard haggling to try to pay less than the bid price for the work. It is worth remembering that an offer only exists until a counter-offer replaces it. Thus a buyer will jeopardize its rights to accept the low bid if it offers to accept the contract at a suggested lower price or under different circumstances to those set out in the bid documents.

Pre-award discussions can cover a broad range of issues. If they have any intent to modify an original offer, then the result is a new offer outside the terms and conditions of the original competitive terms.

A common topic for discussion in pre-award meetings is the price of, or the schedule for, the contract. Sometimes it pays to have cost- or schedule-reduction ideas prepared for these meetings – whether we are the buyer or the seller.

We need to carefully consider the ethics of any added steps that mess with the conventional bid process after the prices have been received, as there is an issue of wasting bidders' time and money if the bid process was not conducted in good faith. Any subsequent negotiation should also be conducted in good faith.

There is a fine line between good business practice and unethical behavior in how pre-award negotiations are conducted.

4.7 Award: Do we Really have an Agreement?
In other work, I have found that there is only one fundamental underlying cause of failure for projects. This is a breakdown in communication. It

may be deliberate, but – most likely – it is not. When communication breaks down, someone's expectations fall out of line with what will happen. This leads to disappointment and then to a loss of trust. Once trust is breached, we see more serious effects that can ultimately lead to disputes and then to litigation. What we want to do in good contract administration is recognize the early warning signs of trouble and then do what we can to head off any problems before they become too difficult to resolve.

Communication is the only cause of failure in a project (Hartman 2000).

When the rubber hits the road we can start to discover some of the things that can readily lead to conflict. Some of the more common ones are discussed in the following paragraphs. We need to look for the early indicators of misalignment of expectations between buyer and seller. This needs to be done before, or at the latest during, the selection process.

Delayed progress payments: Payment is made later than stipulated in the contract. If the payee does not formally register a complaint, this may be construed as acceptance of a modified term of the Contract that allows for such delayed payments.

Changes that are identified as such by the vendor and are not accepted as such by the buyer or its representative end up as claims and may lead to a lot of acrimony and additional cost.

Extra work to the Contract is much the same as a change and may have been assumed to be included by implication (for example, how can an electric door operate without a power supply, even if one is not shown on the drawings?)

Cooperation between the buyer and its vendors may be necessary and may not have been fully articulated in the contract. Assumptions about work sequence, access to the workplace, and overall timing may not have been thought through. Or perhaps this work is being done in an existing office or factory that continues to operate and the assumption was that the contractor would work quietly, not use vibrating equipment, or only work at night or on weekends.

The real value of, and respect for, expertise and the positions of people and their organizations is not always demonstrated and this can leave people disenchanted or even aggressively obstructive. This can lead to disputes and then we are again in trouble.

There are many other ways that we can re-interpret, misinterpret, or simply read something from a different perspective. Any of these normal human habits have the potential to add confusion and even possibly lead to real or implied changes to the terms of the contract. Arguably, signing a contract is not the end of negotiations, but the beginning of a relationship that will require ongoing negotiation and relationship building to work properly.

The value of a written agreement as the basis for a contract lies as much in how it is managed and interpreted (an issue of relationships between the parties) as it does in how it is crafted in the first place.

Contract management is the second hardest part of administering a contract. How effective it is depends to a certain extent on how the contract came together in the first place. There has been a lot of work done by the Construction Industry Institute (CII) and others on the improvements that may be achieved through better contract formation and management. The content of a contract is explored in a bit more detail in Chapter 5.

Chapter 5

THOU SHALT NOT BE
AMBIGUOUS AND VAGUE

> *Clarity in wording: ambiguity, latent disputes*
> *and interpretation of contract clauses.*

In a study of the understanding by participants of the risk allocation implicit in contract clauses, it was discovered that the real interpretation and understanding of the meaning and implications of a contract clause can vary quite widely. This phenomenon is not readily apparent, because we do not normally check for this explicitly. If we ask someone if they understand the contract, they typically say yes or no. If the answer is yes, we assume they have the same understanding that we do. And we will probably be wrong. If they say no, there is a better chance that we will end up with the same (or at least similar) understandings if one party correctly explains the clause and its implications clearly and unambiguously to the other.

In Chapter 5, we present the results of a study (Hartman et al., 1997) published by the American Society of Civil Engineers (ASCE). The implications of a "latent dispute" and how to avoid such circumstances are discussed.

5.1 Language and Technology Growth

In today's rapidly changing world, we are adding many new artefacts, tools, ideas and products everyday. Some experts assess the growth of such artefacts to equate to doubling the number of things we have every three years. While this is going on, the growth of our language is struggling to keep up. Rather than invent numerous new words, we recycle old ones. Effectively, this means that we are increasing the opportunity to

179

misunderstand each other. This is important to understand in the world of contracts in which the written word and its interpretation are the foundation of our actions and responses to situations.

 Recycled words are causing us grief, as they are less clear each time we re-use them.

I TOLD you "Semi Formal" did not mean this.

There are many words in the English language that mean different things in a particular context. Add to this our own personal paradigms, the odd misconception, or misuse, and we quickly approach the challenges faced by the construction crew on the Tower of Babel. The extent to which this is a problem is hard to measure explicitly. To appreciate the significance of the issue, we should look at a few examples.

Take the word "function". What is our function? Who was at the function last night? Which keyboards have function buttons? X is a function of Y . . . and, yes, there are more ways in which this word can function.

In a language in which "fat chance" and "slim chance" mean the same thing, and flammable and inflammable are not opposites, we can expect potential communication problems, so we should continuously be on the look out for them. If this were not bad enough, we tend to be casual with the language and cavalier in our specifications. Most of us have seen specifications that do little or nothing to clarify the intent of the buyer. Again, some generic examples help to set the scene (see You Have to be Joking).

 Common language is remarkably uncommon.

You Have to be Joking

Here are some of my favourite quotes from engineering and architectural drawings and specifications. Hard to believe, these are from real projects.

"The dimensions on this drawing are purely speculative. The contractor shall include for all materials required by the engineer."

"The electrical contractor shall provide for all demolition, both shown and not shown in the drawings."

(Cont.)

You Have to be Joking (cont.)

"The contractor shall mix the three types of brick in the wall in a random fashion to the satisfaction of the Architect. Upon completion of the wall, the Architect shall inspect the finished wall. If not satisfied, the contractor shall demolish and re-build the wall." [This for a wall that was half a city block long and four stories high.]

Some of the more common phrases that leave interpretation open include:

"... to the satisfaction of the engineer (architect)"
"... or similar approved product"
"... or equal"
"... as generally required by the customer (or buyer or client ...)"
"... including but not limited to ..."

How does one price such a specification? What is the hidden agenda? Does this mean the buyer does not really know what he or she wants? Who decides when something complies with this specification? When do they decide? Who takes the risk?

Another problem is the need to use jargon. Jargon is a language that is particular to a very small and specialized group. Nowadays we see included in this the use of three-letter acronyms. A more understandable acronym is PVC (for polyvinyl chloride). Less well known is a PLC (for programmable local controller). Other acronyms and specialized words are more obscure. Yet others are misused, leading to confusion. And then, just to make life more interesting, we have terms that are unique to a company or have a specific and special meaning in one organization and a different meaning in another.

Superfluous words also creep into specifications and into contract terms and conditions. These cause grief if they are not consistent or if they serve no useful purpose. In the latter case, the best we can hope for is that there is more material to read to find out what the customer really wants, including a "definitions" and/or "abbreviations" section in the contract documents. At worst, it may contradict something stated elsewhere and either cause confusion or be misleading.

Jargon can obfuscate as much as, or more than, it can clarify a situation.

The idea of a contract term or condition that is not really clear is not new. We know this from the large number of cases that have been through the courts over ambiguous language and the resulting confusion. Generally, in common law jurisdictions, the courts do not directly interpret the meaning of the contract. The courts generally favor the reasonable interpretation of the party that did NOT write the contract words that are under dispute. The rationale for this is quite reasonable: it is incumbent on the author of a contract to avoid ambiguity in the document that he or she produces. After all, it is the author who knows what he or she wants, not the reader.

The law tends to side with the person who had to understand what someone else said or wrote. The author must be aware of the responsibility to be clear and unambiguous.

5.2 The Implicit Specification

There are some specifications that are implicit in the normal course of doing business. For example, we expect that, when we buy food, it is edible. Similarly, we can reasonably expect an engineer to be competent in engineering or a dentist to know what he or she is doing. This basic concept holds true in most jurisdictions and is often supported by legislation to protect the consumer. There are other factors, too, that set the standard for goods and services. We expect, as a minimum, that what we pay for will at least do what it was represented to be capable of. For example, if we buy a car, we expect it to work. Furthermore, if we purchased or it had included a warranty, we do not expect to have to fight over the necessary repairs should the car not perform, provided that the warranty terms cover the problem. Where problems often arise is in the implied warranty that we may be given and that may or may not be either real or honest. Here is a common example:

We are considering buying a secondhand car from a dealer. The dealer says that the car we want was checked out in their service shop and that there is nothing wrong with it. The dealership is for a well-known, quality import car. We rely on their integrity and buy the car. It falls apart within the first six weeks. The dealership says that this is our problem because as far as they are concerned, we purchased the car "as is". What should we do? Have we been treated fairly? How much of this is dependent on local custom and what may be considered fair practice? What is our best recourse? These are not easy questions to answer, as they are dependent on fine detail, local traditions, what the letter of the law says in our jurisdiction, our personal style (e.g., aggressive or passive) and many other factors. The very same issues exist in other situations.

Consider one of these other situations. We have hired a contractor to repaint a boardroom. The job is done and we are presented with an invoice. The color is wrong, however, so we ask for it to be repainted. We specified a color called "teal blue" based on a chart supplied by a paint manufacturer, but the paint used is too green. The painting contractor pulls out a chart from another paint company and shows us the color "teal blue," which exactly matches the color the contractor used. Now, go back to the questions in the previous paragraph. Repeat the exercise for a cracked concrete slab. Or a piece of equipment that does *almost* everything it was expected to do . . . If the wording in the agreement with the vendor has the terms "common practice" or "good practice" or "acceptable standard" in there, we are probably in a worse situation than if it did not because we now have two arguments. First, what does "good practice" really mean? And second, does what we received comply with this definition?

"Common practice" and "good practice" can be dangerous terms to use, as they may vary from one location to another. In any event, they are hard to accurately define.

Even if we have explicit specifications, there are also implicit ones. Public safety, compliance with all legal requirements and not breaking the law are obvious ones. In the absence of any specification to the contrary, what is delivered should be fit for the intended purpose, provided that the intended purpose is clear from the outset.

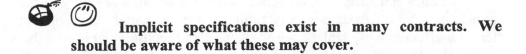

 Implicit specifications exist in many contracts. We should be aware of what these may cover.

Again, there are some really basic implied conditions that normally exist (e.g., safety). The law of tort in common law jurisdictions typically protects third parties from the fallout of another party's activities and *vice versa*. We cannot contract out of this, but we can assign responsibility for managing the situation and dealing with failure to manage it adequately to one of the parties to a contract. Although anyone can sue anyone else in tort, there are opportunities to manage this within the structure of the law in most jurisdictions. The real question is whether it makes good commercial sense to do so.

 We cannot contract out of liability in tort. Safety of employees and third parties is a requirement in law in most jurisdictions.

It is safe to assume that the intent of a legally binding contract is basically legal in itself. Thus, we can normally assume that a contract will not require us to break the law. What may this really mean to us? We contract to a client that we will deliver a product to a given destination within a specified time. It is not possible to deliver in that time if we drive safely and within the speed limit. Is this contract enforceable in law? I would argue that it is. Unfortunately, contracts do not have to be easy or even fair. We contracted to do something that could be done, legally, at great cost (e.g., use a helicopter). The contract is, therefore, arguably enforceable, as the method of working is normally left to the vendor unless it is explicitly set out in the contract.

Illegal activities are never part of a legally binding contract. We cannot be required to perform an illegal act to comply with a contract, but this does not relieve us from the obligation of finding a more expensive, but legal, solution.

Codes and standards, especially when the health and safety of the public may be affected, normally form an implicit part of the contract terms and conditions. Unless we are aware of specific exemptions that apply to a contract, there is an implicit expectation and obligation to comply with all relevant regulations, whether we know they exist or not. It is our responsibility to find out just what those regulations are.

Normally, the codes of practice, standards and other published guidelines that affect work under a contract are effectively implied in it.

Professionalism carries responsibility. There is an expectation from the general public that all professionals maintain certain standards. In the case of regulated professions (medicine, law, engineering, etc.), there are minimum standards of competence and expectations that we will work with in our area of expertise. There are also expectations that we will disclose risks to the buyer when these may be reasonably known to a professional. Finally, although this is not a comprehensive list, we need to be aware of what "best practices" may mean for the professional. Professionals are expected to stay current with their profession. If we offer professional services, then we are expected to deliver our services in line with good common practice at the time. Failure to do so may mean subsequent liability for any of the fallout from shortcomings in the product that was the result of our professional service or was directly affected by such a service.

When professional services are provided, the standards associated with that profession are implied in the contract.

Craftsmanship and the resulting quality of workmanship are often implied expectations in a contract. It used to be that the quality of workmanship from a trade was that which could be obtained from someone who had undertaken a full traditional apprenticeship or its equivalent. With the effective demise of this old approach, new standards are being created. These reflect new ways of working, newer materials and technologies and often a degree of factory assembly or pre-assembly. The implied standard is always difficult to enforce, as it is difficult to define with any accuracy.

Minimum standards of craftsmanship or workmanship may reasonably be expected. What constitutes acceptable standards may be harder to define, especially if there is no basis for such a definition.

Cultural differences, business practices and other variables also affect expectations, which vary in different parts of the country or in different countries. If the buyer and the vendor have different expectations for these implied specifications, then we have a latent dispute. This is but one kind of latent dispute.

There may be other expectations implied in how a buyer or a vendor normally does business.

5.3 Latent Disputes

Latent disputes are the disputes that lurk in our contracts and rear their ugly heads only when some event triggers them. They are in our Agreements because we have interpreted them differently, and this difference does not become apparent until something goes wrong – or at least does not go as expected. I do not include in this a potential dispute that could not have been detected earlier. So a key part of the definition is that the dispute is potentially detectable. The trick in managing these is in detecting them in the first place.

A latent dispute is one that could have been identified earlier than when it occurred.

Generating a Problem

The owner of a facility needed an emergency generator added to part of an expansion project. This requirement came as an interpretation of a bylaw that became effective once the facility exceeded a certain size. The owner of the facility decided to add a piece to the bid documents requesting a separate price for the supply and installation of this emergency generator. There were two reasons for this. First, the owner still had a vague hope of persuading the local regulatory agency that this generator may not be needed after all, and second, he wanted to see if he could find a used unit that may be cheaper than the one that the low bidder offered. The wording in the bid documents was clear. It stated, "The bidder shall provide a separate price for the supply and installation of an emergency generator as specified in the drawings and specifications that form part of this set of documents."

The base bid for the low bidder was about $11 million. The separate price for the generator set was $2 million. The contract was awarded. The owner found a cheaper generator and issued a change order for a credit against the $11 million in the amount of $2 million. This reduced the value of the contract to about $9 million.

The contractor was shocked. He refused to sign the change order, pointing to the bid documents. The wording was clear, the owner had asked for a separate price. If he wanted the generator supplied and installed, he would have had to issue a change order for an additional $2 million. This would have increased the contract price to about $13 million. The outcome: a dispute!

The contractor reluctantly agreed to install the generator while this dispute was being resolved. Both parties had agreed to go to mediation. One of the first things that the mediator did was point out that the problem was not that simple. The separate price quoted was for supply and installation. Now the $2 million needed to be split into two components. How much was for supply (now being done by the owner), and how much was for installation (now being done by the contractor)?

The different interpretations of what a "separate price" meant went undetected until the event that triggered application of the two different meanings. This is a classic example of a latent dispute.

When working with Pat Snelgrove, he and I came across the concept of a latent dispute during an interesting piece of research. Originally, this research was to measure the effect of changing an industry-standard contract. In this case, it was a Canadian stipulated price contract standard known as CCDC 2 (Canadian Construction Documents Committee 2). This well-used document was re-issued after careful review in the mid-1990s. We sought people's opinion regarding the shift in risk implied in the contract. We found very little common agreement in the interpretation of who carries what risk in either the new or the old contract clauses. This surprising result got us thinking about the concept of a latent dispute. . .

Latent disputes are created by initial misunderstanding or miscommunication.

Now that we know what a latent dispute is, we can readily understand the human element of what happens with a particular interpretation of a contract clause. Over time, and as the interpretation we have is not challenged, we are more likely to believe and accept our view of what the contract says on a particular point. The longer we leave a clause with a latent dispute, the harder it is to fix and manage expectations when we have to address the problem later.

A latent dispute is made potentially worse by hardening of misaligned expectations over time.

5.4 Risk as a Tool to Identify Latent Disputes
On the basis that latent disputes have a possibility of turning into real ones, how do we know when the best time is to address the dispute? For most people, knowing where these potential problems are is, at least, part of the solution. We also know that when there is ambiguity in the contract, the authoring party is most likely to be held responsible for the ambiguity. Remember, courts generally find in favor of the party that is not the author.

It would seem pretty obvious that it is in the interests of the party that authored the agreement to make sure that latent disputes are ironed out at the beginning. The author is usually the buyer or its representative.

As a buyer, there is maximum negotiating leverage on any changes in pricing or other advantage that may fall out of the process just before the contract is signed. Most buyers know this. What many buyers forget is that this stage of the negotiation helps establish the relationship between buyer and seller for the rest of the contract. If that relationship is based on a feeling that the seller has been taken advantage of, then there may be some damage to the level of performance that will follow. It will do nothing to reduce subsequent stress levels.

Identifying all of the latent disputes will significantly reduce stress and subsequent problems on projects.

Latent disputes will not be identified if we ask questions along the lines of, "Do you know what this clause means?" The answer to such a question is unlikely to be anything other than "Yes" simply because anyone who has read a clause assumes that their interpretation is the correct one. We need to be able to ask the right questions to find out about differences in interpretation of a clause. A good start is to discuss what would trigger looking at that clause, what the implications would be, and what would happen if someone were to enforce it.

Part of the process of identifying a latent dispute is asking the right questions.

Asking the wrong question does not help find a latent dispute. And the obvious question is the wrong one!

In a questionnaire that we administered, the question that gave us the first clue about latent risk was the following, "Who has the risk?"

Simple question. The interesting result was that people who had been interpreting the meaning of the clauses about which we asked this question had widely differing assessments! The respondents to the questionnaire included contract administrators, project managers, claims specialists and lawyers. No two people in the same organization or on the same project team had the same interpretation, measured in terms of who carries the associated risk, of the clauses in the test contracts. Almost all of the respondents had two things in common. First, they were frequent and experienced users of the test contract. Second, they were surprised that there were such different interpretations of the same clause.

 The right question to ask is, "Who really has the risk associated with this clause?"

The CCDC 2 document we used for this research is a Canadian standard stipulated price contract, used extensively in industrial, commercial and institutional construction projects. It was in the process of being replaced with an updated version. Originally, we were interested in assessing the difference between an established version of an industry standard and a new version. We were interested in a number of aspects, including the level of acceptance of the new document relative to the old one, differences in perception of where the risk lies in the two documents, perceptions of bias and shifts in the two versions and whether one was easier to interpret and understand than the other. One of the tools we developed to test some of these aspects was a Lickert scale, which we used to obtain respondents' views of who carried the risk associated with particular clauses. It was this scale that disclosed the big differences in opinion between the different experts regarding what these clauses really meant.

These differences have been reported in several publications (Hartman and Snelgrove, 1996; Hartman et al., 1997; Jergeas and Hartman, 1996).

 The concept of a latent dispute arose out of research into interpretation of "well-understood" contracts.

Who owns which risks?

 Risk should lie with the party best placed to handle it by managing or absorbing that risk.

Oh, What a Brick!
(See also You Have to be Joking)

We were building a major extension to a significant cultural center. Part of this project included a large blank brick wall facing onto a public park. The wall was half a city block long and four stories high. The architectural specifications required three different colors of brick to be used in a specific ratio. Finding out the numbers of each type of brick was easy. But the specifications also required that the bricks be placed in a random fashion, to the satisfaction of the Architect. So I had to ask, "What will satisfy the Architect?" The reply was simple, "Build the wall and then we'll tell you if we like it!" Not wanting to waste time and money building a wall of that size and then having someone decide they did not like it, we agreed to build a "sample" wall, six feet wide and six feet tall. The real wall simply needed to look similar to the sample.

The words "to the satisfaction of the Architect" hung over our heads until the wall was accepted, as we would have had a dispute of some kind had the architect rejected our tradesmen's interpretation of what would look like the sample wall. We had a latent dispute resulting from a lack of clarity in the specifications.

So what does a latent dispute look like? It may relate to a contract bid and the expectations around that bid (see Generating a Problem in Section 5.3). Here are a few other examples:

Ambiguity in specifications or a lack of clarity leads to a potential future dispute (see You Have to be Joking). A contractor can only hope to guess at the implications of such contract terms or specifications.

Contract wording can also obscure the scope of the work. The simplest example of such wording, from a pragmatic point of view, is the use of the following three elements in a contract in combination. These three elements are commonly used together:

- The phrase "time is of the essence" inserted in the contract, usually in the General Conditions
- A Stipulated Price Contract
- Rigorous specifications and no flexibility in delivery (such as use of alternatives).

What this does for us is eliminate – or at best significantly restrict – any opportunity to work around problems as they arise during the execution of the contract. We have constrained the contractor in terms of time, money and performance. These three elements need to be prioritized for effective project management (Hartman 2000). In the absence of any flexibility, the contractor only has the option to be creative in managing any situation that presents itself to recover any losses. The three elements of time, money and performance work together (see Just a Matter of Time).

Just a Matter of Time

As engineering got underway on this design/build project, it quickly became apparent that the client was unsure of exactly what he wanted. The senior engineer who was charged with working with the client was still young and had worked primarily in design, with little site or construction experience. She was in this role because of her known skill in managing people and their expectations.

As design progressed, the client's representative asked this senior engineer what the effect would be if they were to request installation of a different valve in one of the process pipelines. The senior engineer checked and was assured that, because the design was still at a preliminary stage, the effect would be negligible if the client made that decision quickly.

(Cont.)

Just a Matter of Time (cont.)

This created a latent dispute. The client's timescale for "quickly" was much longer, measurable in months, compared with that of the designer, who was thinking in terms of hours. The reason for this was simple: this was a design/build project. It was being fast-tracked and key long-delivery items (which included the valve in the original design) were being ordered as specific parts of the project were detailed.

When, some two months later, the client requested the change in valve, he was shocked when the senior engineer requested a formal change order and pointed out that the original valve had been ordered and the pipe spool fabricated and delivered to the site, ready for installation. The change being requested now would delay the project by months and would include the cost of the new valve, as the old one could not be returned. In addition, there were costs associated with disruption of the construction schedule to work around this issue and the cost of installing and removing the valve that had originally been specified and accepted by the client when the contract was signed.

In this case, a latent dispute arose over a change and some miscommunication resulting from the specific definition of the word "quickly". The solution on this project was to build to the original contracted specifications and then the client could make the change after the project was complete and accepted by the operators. Resentment remained, however, over what the client saw as gouging by the contractor. The contractor saw it as a lack of knowledge ("stupidity" was the term used) by the owner.

It is common practice to include clauses in contracts that address the impact of delays to the work. This work may be defined explicitly, or it may be implicit. Either way, the significance of the clause and how it is interpreted could lead to creation of a latent dispute. There are many other examples of potential latent disputes, but I have picked this one because delays are so common on projects.

Delay clauses in contracts take many forms. The following is one set of clauses that address delays (Goldsmith, 1993).

GC 4   DELAYS

4.1   If the Contractor is delayed in the performance of the Work by an act or omission of the Owner, Consultant, Other Contractor, or anyone else employed or engaged by them directly or indirectly contrary to the provisions of the Contract Documents, then the Contract Time shall be extended for such reasonable time as the Consultant may decide in consultation with the Contractor. The Contractor shall be reimbursed by the Owner for reasonable costs incurred by the Contractor as the result of such delay.

4.2   If the Contractor is delayed in the performance of the Work by a stop work order issued by the court or other public authority and providing that such order was not issued as the result of an act or fault of the Contractor or anyone employed or engaged by him directly or indirectly, then the Contract Time shall be extended for such reasonable time as the Consultant may decide in consultation with the Contractor. The Contractor shall be reimbursed by the Owner for reasonable costs incurred by the Contractor as the result of such delay.

In addition to these two clauses, there are others that deal with strikes, written notice of delay within 14 days, the schedule made by the Contractor, and delays incurred by the Consultant.

These are equitable terms for addressing delays. However, when we asked contractors and owners to say who carried the risk associated with delays under these terms, owners generally felt they did and contractors generally believed they were at risk. Why would this be so? The primary reason lies in the wording and its interpretation. The clause is sprinkled with words and phrases such as "reasonable", "directly or indirectly" and "may decide". This terminology raises the level of anxiety of people trying to determine where the risk lies. These words and others leave the clauses somewhat open to interpretation depending on many situational factors. One important factor is the person charged with administration of the contract. Many owners and contractors had memories of such clauses in which the interpretation at the time was seen as biased against them. (It seems to be easier to remember the unpleasant events rather than the others in which we really got what we rightfully deserved.)

This situation is one that carries a latent dispute.

To help find such latent disputes and to start to reduce their likely future impact, we turned a research instrument into a working tool. We call it the Risk Ownership Scale (ROS). This scale can be reproduced on a rubber stamp. It can be used simply by scoring each clause, using a scale from –5 to +5. The range represents full ownership of all the risks associated with a clause or subclause by the Contractor at one end and the Owner at the other. Figure 5-1 shows how it might look.

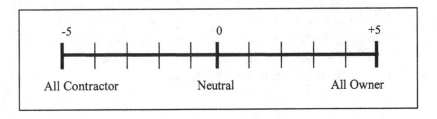

Figure 5-1. Risk Ownership Scale.

We use this scale against each clause that the Owner or the Contractor (or the Consultant) feels may be of future concern in administration of the contract. Each party places a mark on the scale, using a different symbol or color to represent where they see the risk. If the risk marks do not coincide, there clearly is a difference in the interpretation of what that clause means. This difference, now that it has been identified, can be discussed and resolved while everyone is level headed and before someone has a particular position in a real situation on the project, when true objectivity may be harder to achieve.

The purpose of the Risk Ownership Scale is to help start an open discussion regarding who will accept responsibility for a particular risk.

Although use of the ROS helps identify latent disputes, this is not enough. Now that we have identified the problem, we still need to resolve it. This is most easily done by clarifying the intent and making whatever adjustments to the contract that may be needed to address a perceived change in the terms and conditions of the contract. There are some risks in this approach as, like so many other things in the contracting business, it is open to abuse. The lowest risk approach is to use the ROS before going out to bid on the work.

Just using the Risk Ownership Scale is not enough. Follow through with a decision, allocate risk clearly and adjust the price for the contract, if necessary.

5.5   Clarification of Clauses: Before or After Bid?

We have just discussed the use of the ROS and suggested that it is best used before going to bid. We will now look at this recommendation and see why it was made. Allocation of risk is just one of the issues that warrant clarification before signing a contract. Most practitioners who deal with contracts and their award are familiar with the myriad of issues that arise and are unique to each situation. As buyers, we worry that the lowest price has, in fact, not included for everything. Our concern may be stimulated by anxiety over performance or survival of the contractor or we may believe that there will be a lot of problems stemming from an underpriced offer to do the work or supply something. From a vendor's perspective, we worry that too much money has been left on the table or that there was a significant error in our bid.

Clarification of a bid addresses the issues that become apparent during the bid process. Whether or not clarification is sought before or after submitting a bid is often a decision that is based more on a business assessment of the situation rather than on simply obtaining clarity. For example, if we see ambiguity in a set of bid documents, we may wish to interpret the document in the way that is most favorable to being able to submit a lower price, as that may be what gets us to the negotiation table. Equally, if there is an opportunity for the competition to underprice us because we know something that the other bidders may not, then it may be more important to obtain a clarification that puts us back on a level playing field. Generally, for an owner, it pays to have as much clarity as possible from the outset. That said, there are cases in which some information has been withheld or modified by buyers in the hope that its absence will lead to lower pricing. In one case, a soil investigation was suppressed, as it identified groundwater problems that had been missed in a simpler study completed the previous year. Only the earlier study was provided to contractors bidding to build a factory on that site. In another case, information on the availability of materials was falsified to encourage bidders to come up with lower prices.

Clarification of contract clauses before signing a contract is important. Whether or not a clause is clarified before or after bidding seems to be more of a business decision.

I have appeared more lenient toward "games" played by a contractor than ones played by the owner in the bid process. Why? One reason is that buyers set the rules, as they should, for how they will select the contractor. For fairness (the underlying issue in most cases), all bidders at least need to be on the same playing field, whether it is level or not. There are few opportunities for a contractor to offer an advantage and discuss the project unless it can get to the negotiation table to finalize a contract. The magic for the seller, therefore, is to get to the table. The rules are different if the owner sets out to negotiate without first looking for the lowest price. But this is within the control of the buyer. Any implied harshness towards the buyer lies in the relatively more powerful position of the buyer in any acquisition situation – all other things being equal. In the end, though, if we strike a deal based on such games by either party, we will probably increase our chance of failure. It is important that the final deal is clear, as simple and straightforward as possible and based on honest negotiation.

What are the relative merits of clarifying before or after bidding?

One small typo and we have to provide Finnish dates.

Clarification before the bid has both advantages and disadvantages from either the buyer or the seller perspective.

What would happen if we were a contractor bidding on a project and we saw a problem in the bid documents that we suspected was not obvious to any of the other bidders? This may be because we are familiar with the client or have a new technology that addresses this problem or else because we have had experience with a similar situation. This problem creates an opportunity for us to submit a more competitive price. Either we rely on the change we believe to be inevitable later or we offer a solution as part of the bid. Does it pay for us to alert the buyer and lose this advantage?

Contractors who identify a problem or opportunity before submitting a bid may consider that doing so equates to giving up a competitive advantage.

In the same situation, we see the problem and have a solution but have no particular advantage. Yet we know that the issue will alarm the other bidders (we hope all of them!) into overpricing the perceived risk. Will we be tempted to raise the concern – even laying it on thick – so that the competition may be encouraged to overprice a potential and maybe half-considered solution? Will we be tempted to time it so the competition did not have enough time to investigate the problem properly? Will we want to strongly encourage the buyer to ensure that the other bidders were aware of the problem?

Some questions are worth asking, from a contractor's perspective, if the expected answer will scare the competition into adding premiums to their bids, thus increasing our chance of success.

If we are a buyer and we know exactly what we want, do we want to have a contractor offer an alternative that someone in our organization may find appealing because it is cheaper in the short term? Have we not just spent a lot of time developing the solution we have? Are clarity and precision in a bid not a good thing? Will this not make life easier later when we have to manage the contract? The answer is yes. But. The "but" is here because there is a trade-off to be considered. Some suppliers and contractors have a lot of experience in doing what they do. They may have seen our solution fail recently for a reason that is clear to them but not obvious to us. They may have seen our competition come up with a much better solution. They may have been exposed to a technology we are unaware of. All of these opportunities will be missed if we do not leave a door open to allow for ideas, innovation and other potential value-add opportunities that the contractor or supplier may be able to offer.

He did say he was a "heart plumber",
but I thought he was just being modest.

 Absolute clarity in a set of bid documents may eliminate contractor creativity and, therefore, opportunities for the buyer.

If a problem comes to light in the bid process, it is good practice, and is mandated in some organizations, that the issues be shared with all bidders. Some practices require all questions from any bidder and the corresponding answers to be distributed to every bidder. This is a good idea, as long as it does not inhibit a bidder from asking a question.

Any issues or problems raised before a bid closes can be shared with all bidders and the issue can be eliminated before it creates problems in bidder selection or subsequent management of the contract.

What are the relative advantages and disadvantages of clarifying after the bid?

One of the biggest disappointments we can have as a vendor is to be the low bidder and lose the work. We may be better off without the work, because we have missed a critical component in our price. (Does this ever happen?!) We may feel that we have to absorb extra costs, eroding our margin if the buyer seeks clarification and we find they are asking if we included things we did not. This may make the project less attractive.

The contractor may have to adjust its price in response to a contract clarification and this may make it uncompetitive. Alternatively, it may have to absorb any additional cost associated with a clarification.

Equally, if the contractor has a strong indication that it was low by a significant margin, it may wish to adjust its price upwards at the expense of the buyer. One company I was exposed to for a while used to bid only on projects that had a public bid opening. It was always the low bidder. Within seconds of all the prices being announced, it would withdraw its bid, claiming that it had made a mistake. It would then resubmit a price that was still low – but only just. This practice stopped after landmark litigation closed the door for this practice in Canada, where this company operated. Some buyers require a bid bond to manage this type of behavior. Less overt is the practice of adding money for missed items in the pre-award discussions until a bidder feels it has closed the gap between the next lowest bidder and its price.

If the contractor knows it has a competitive proposal, clarifications may create opportunities to negotiate a more advantageous deal.

On the other hand, if a bidder knows it did not submit the lowest bid, will it consider reducing its margins, renegotiating with key subcontractors and suppliers, or taking other steps that will give it opportunities to work the price down in a clarification meeting?

 If the contractor does not have the best proposal, clarifications can create opportunities to improve its competitive position.

We have submitted a bid. Now it appears that the post-bid clarifications are exposing significant ambiguities that may challenge our proposal. What we thought was a clear interpretation is now being challenged by the questions asked by the buyer. There are a growing number of points that appear to be open to interpretation and each one adds to our risk or to the cost of doing the work. What should we do if we were initially the low bidder?

 The bids are in. If there is any doubt or ambiguity, the contractor may request an increase in price or a re-bid if this increase makes it noncompetitive.

The process of clarifying a bid creates opportunities to add minor items to the scope of the work, if they were not clear in the bid document. This is easy to do if the low bidder is in need of the work. Ask questions like, "You DID include for the cost of maintaining all of our site roads during the construction period, didn't you?" or "Snow removal IS included, isn't it?" or "The site office for the client staff and the consultant team – with air conditioning – is included in your price?"

 The owner has an opportunity to negotiate a better deal using contract "clarification" as a vehicle for doing so.

We are looking at what we may arguably consider to be game playing. And now that we know the basic advantages and disadvantages of getting clarification on the meaning of clauses at different times, we can deduce a set of "rules" for the game that is played in the bidding process.

> **Games played in bidding are generally counter-productive in the long term. And they tend to destroy trust – an essential ingredient for both parties if the contract is to be managed effectively after award.**

5.6 Managing Expectations: The Soft Side of Improving Contracting Effectiveness

If you have gone along with my logic to this point, you will be ready for the next obvious question. Maybe it is not so obvious. If we accept that people play games in the process of trying to win work and their trying to actually make some money doing the work, we should ask ourselves how much of what we do is governed by the contracts we write and how much is governed by other factors. And what are those factors? How significant are they?

A few years ago, I read about a group of consultants in the United States who would not do work with anyone who required a contract to be in place. They would put their reputation on the line and, in exchange, would expect to be paid for the value they delivered. Unfortunately, I never noted the details at the time. The concept stayed with me though, as we did more work on the role of trust in contracts and in the more general business of delivering projects. The obviousness (to me at least) of the questions in the previous paragraph is undoubtedly linked to the story of this group of U.S. consultants. Another experience that supports the need to ask questions about the role and nature of contracts is a series of three studies that I had a part in over a ten-year period. All three of these studies were to investigate drivers for repeat business for engineering consulting companies. All had as their primary finding a direct correlation between the consulting engineer understanding and working with their client and repeat business. All also had another interesting finding: No correlation between repeat business and being on time, being within budget, or delivering on scope and quality.

Specifically, these three separate private studies have consistently shown that engineering consulting (design) firms get repeat business because they manage their clients' expectations. The connection was one-to-one with perceptions of success. The perceptions of success were linked to good management of expectations around what could reasonably be achieved.

Repeat business is typically earned through being realistic and managing the expectations and requirements of customers.

What is this telling us? I believe the message is that "soft" issues, such as the relationship between contracting parties, is probably one of the most significant factors in the success of a contract. Now let us take this one step further and look beyond just the Western mindset. In other cultures, such as the Eastern cultures of China and Japan, it would appear that the formal signing of a contract has a different significance to what we perceive in the West. There, this point in the life of a contract is where we have set the ground rules for subsequent negotiations. For the West, this point normally marks the end of the negotiation phase. We know that in reality changes are inevitable, and we end up negotiating these as they arise. Are we fooling ourselves when we place so much reliance on the contract, its legal significance and how we can squeeze every last benefit out of what we are entitled to under its terms? Should we be looking for a softer approach that recognizes that we are human and that it is other humans whom we need to persuade to make changes and achieve success?

 In many Eastern countries, a contract is simply a basis for subsequent negotiation. This is quite rational if we think about it.

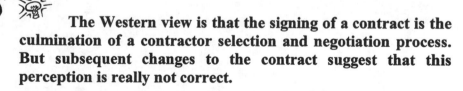 **The Western view is that the signing of a contract is the culmination of a contractor selection and negotiation process. But subsequent changes to the contract suggest that this perception is really not correct.**

 In reality, we make endless changes after contracts are signed. Some are formalized. Others are not. Are we not really using a contract as a basis for subsequent negotiation?

When we negotiate changes, as well as how we approach these negotiations, seems to be critical to the perceived success of the project.

The trick seems to lie in maintaining good relationships. This is achieved in large part through open communication. Part – and probably critical to this – is effective management of the other party's expectations.

Any supplier who is successful is one that delivers on what is promised. This cannot be done if what is promised is impossible.

Buyers do not like surprises, especially bad ones.

We have just looked at a specific risk in preparing contracts: ambiguity and vagueness. Now let us look at risk in a broader sense.

Chapter 6

THOU SHALT SHARE OUT RISKS
EQUITABLY AND WITH INTELLIGENCE

Risk assignment through contracts:
risk assessment, cost and allocation.

One of the prime purposes of a contract is to assign risk. Whether this is risk of execution, payment, working conditions, performance, environment, safety, or anything else, risk is assigned in contracts.

Some common risks that are assigned include timely performance (e.g., through liquidated damages for delay clauses), environmental risk – such as weather or soil conditions (through performance clauses or bid terms) and regulatory compliance (through design and performance clauses). Inappropriate assignment of risk can be expensive. The cost of such risk assignment through exculpatory clauses was the subject of a research project inspired by harsh experience. An example of such experience was described in Chapter 2 (see So You Think You Got a Deal?) The resulting study revealed for the first time the magnitude of the premiums that are associated with exculpatory clauses. Chapter 6 presents the findings of this study. To avoid any scornful laughter, we re-did the study with a different research methodology and sample. The results were essentially the same.

6.1 The Cost of Assigning Risk: The Study, its Limitations and How it was Validated

Risk is one of those unpleasant aspects about doing business that is always with us – like death and taxes. Also like death and taxes, we do not like to think about it and we hope it will not bother us... at least for today. Ideally these things happen to OTHER people. In developing contracting strategies, we have displayed this natural aversion to risk more consistently and profoundly than perhaps we are prepared to accept. The cost of passing risk onto others – which is often what we do – is considerable and perfidious. We have developed habits that mask this cost and we can no longer afford the price we pay for doing business this way. Chapter 6 describes how we investigated the real cost of transferring the risk and what we are doing to reduce the effects of inappropriate risk management practices as displayed in contracts today.

How can you have finished the project?
We are still working out who should take which risks.

6.1.1 The Origins of the Study

I used to run a construction company. Before that, I was a consulting engineer, and before that, I worked as an owner's representative on large projects. In between, I worked on other projects and in organizations that exposed me directly to other forms of delivery of goods and services. A few patterns appeared that struck me as incongruous. Wherever I was in the food chain, the wisest people told me that I needed to beware of the others in the food chain for two reasons. First, they were out to cheat us, and second, they were remarkably inept at what they did. It also appeared that whoever had been involved in the project before had identified risks and had passed them onto the next person in line. Owners passed risk onto consulting engineers or architects. The consultants passed risk onto the contractors. Contractors, in turn, passed the risks onto their subcontractors and suppliers, and so on. There are endless examples of situations in which someone was persuaded to spend money on questionable protection (see Now, Where is the Safety Net?)

Now, Where is the Safety Net?

Once I started collecting stories that we could use to demonstrate additional costs associated with a particular clause in a contract, it became clear that often the Owner got no value for the extra money it paid for some apparent safety net.

In one case, a buyer had included a clause making every supplier of equipment to a petrochemical plant liable for consequential damages arising from the use and failure of their equipment. They did not even consider misuse of the equipment, failure to maintain it properly or any other mitigating condition. One North American supplier was asked to quote on this and they deleted the clause, as they felt it was too high a risk for them. They are one of the largest manufacturers and suppliers of this type of equipment in the world. Their bid, without this clause, was $80 million. Another supplier based in Japan offered to provide the equipment for about $120 million and did not exclude this clause. They were awarded the contract.

The owner had paid a premium of $40 million for the privilege of being able to sue the supplier in the unlikely event that their equipment failed. The jurisdiction for the lawsuit had not been specified. So the probability of actually bringing this suit to a court was diminished. Furthermore, any intelligent supplier from another country would likely have a local company with virtually no assets in place through which such contracts were entered into, protecting the parent from such risks. In the end, the premium was likely totally out of proportion to the real or imagined risk that was being "managed".

As a contractor, I was no different. In my days as a consulting engineer, I was trained to over design (the owner is paying to protect us from failures in the design). The cost of mistrust is inordinately high. But maybe it is worth it if we believe that we are likely to be taken advantage of if we do not include such protection.

Nobody likes risk. The obvious solution is to pass risk onto others.

If we decide to pass risk onto others, this simply transfers liability. The effect is similar to buying insurance unless the other party truly has control over the risk in question. In other words, a contract clause written by a designer on behalf of the buyer that passes any risk of errors in the design to the contractor is probably a bad business step for the buyer. Why? Because the buyer has already paid the designer for a design and is paying a contractor to either validate the design or to assume the risk should there be a problem with it. The risk itself does not disappear. Perhaps more important to consider is the effect on the final cost of the project, as the cost will reflect the premiums that a supplier will charge for carrying the risk. The less reasonable the risk assignment is – all things being equal – the greater the premium is likely to be.

Passing risk to others does not make the risk disappear.

If there is a premium that vendors charge for taking a risk, then what is this premium? At first glance, it may be argued that contractors competing on price will not ask for a premium, as this would make them non-competitive. Reality seems to lie closer to the existence of a premium, either in the form of a direct cost or else in the added process or defensive behaviors that we exhibit in the face of such added risk. If this turns out to be a real cost, does the elimination of such cost with no real value added constitute an opportunity? It seemed, a few years ago, that this was worth investigating in a scientific way and based on real information from the practitioner community.

The big question is whether the tendency to reassign risk really costs us anything. Are we missing out on opportunities?

Before we get to a quick look at the investigation, it is worth understanding the environment in which we work. In Section 6.1.2, we address this in terms of some of the more widely held beliefs in the business world.

6.1.2 Common Beliefs

One of the most obvious reasons for entering into a formal contract for the acquisition of goods or services is to provide specific protection from the unscrupulous behaviors of vendors. Many (most?) vendors are not unscrupulous. The trouble is we do not necessarily know which ones fall under one category or the other. The safe option is to assume all are unscrupulous and are "out to get us". As a buyer, we like to feel that we can be protected from any vendor who may try to take advantage of us. We know that there is legal protection through common law and legislation, but we usually have some very specific requirements or concerns and want to establish our position and rights before entering into a legally binding commitment with the vendor.

Buyers of goods and services have a right to protection from unscrupulous behavior of their vendors.

It is the buyer who normally sets the terms and conditions of a contract. The most common mechanism is to ask a consultant or a lawyer to prepare the documentation. The professionals involved in the development of such documents rarely build them from scratch. The document is developed from past similar contracts and is modified to address specific issues and concerns related to the current contract situation. In addition, in the interest of minimizing risk to the buyer, clauses are modified or added that address more recent experiences. The mandate, stipulated or implied, of these professionals is to protect the buyer. There is rarely any reference to the business efficacy of the protection being offered. Are we asking for protection in situations in which the cost of that protection exceeds the risked probability and the cost or other impact of what we are defending against?

 There appears to be some confusion in the understanding of what is fair protection from unscrupulous behavior and what is a good business risk.

In the end, we are really protecting ourselves against the bad behavior of people. Although we do a lot of business in which one company contracts with another, it is the people who administer and manage the contract that affect how the deal really works. Experienced contractors try to determine how the people who are administering the contract will behave. If they have a track record of fairness, the risk in doing business is reduced. What is this telling us? Relationships between the people involved in working together under a contract will likely play an important part in the success and effectiveness of a particular deal.

We really only work with humans in business. Relationships, attitudes and behaviors are an important and integral part of our business.

If relationships are important and influential in the outcome of a contract, then what should we be looking at changing in our contracting processes to improve the business effectiveness and, therefore, reduce the total cost of procuring goods and services?

At this point, it is worth remembering the Zen tranquility prayer: "Change the things that can be changed, accept those that cannot and have the wisdom to know the difference."

We tend to feel that most of the practices in contracting – from both the buyer and seller perspectives – are hard or impossible to change. Perhaps we really need some fresh thinking. Maybe we need to try to change things that our conventional wisdom tells us cannot be changed. To understand the cost of poor or inappropriate risk assignment, we undertook a study to see if we could identify and measure this cost.

There is a lot to be said for the implications of the Zen tranquility prayer, but first we need to know what we can change.

6.1.3 The Investigation

The ultimate genesis of this study was a bald statement made by a gruff manager of a construction company. He said no contractor ever considered risk when bidding, as doing so would render the company non-competitive. Specifically, he said that in bidding on a stipulated price contract, it was the lowest price that won. It was impossible to be the low bidder and to have covered any of the risks with a premium of any kind. At the time, there was a ring of truth in what he said, although my instinct screamed at me that this was not sound thinking. Why? The answer lies in the following simple logic: if we are a successful contractor, we remain in business because our total income generally exceeds our total costs. Part of our cost is paying for the risks that happen in our work. Where does the money to pay for this come from? It is not normally insurable risks that cause a problem, so it does not come from insurers. If we make a profit, then it does not come from reduced margins. Barring other businesses that support a loss-making contracting business, the only source is the contractor's clients. So how does the contractor collect the premium if typically it is unaware of the fact it is doing so? In our study, we needed to find the sources of revenue that constituted the premium. To do this, we needed to identify the cause and the effect. So we started to look for the possible causes of additional risk in contracts associated with exculpatory clauses.

First we had to find the possible sources of added cost in the contracting process.

Finding the things that add cost to a contract – specifically the items that add cost to the initial price that a contractor proposes – is not easy when self-defence and survival habits have evolved over a long time and are deeply ingrained. Because they are so much part of what we do every day, we see the steps we take to protect our businesses and jobs as simply normal practices. It is, after all, what we do for a living, and we do not have the time to stop and meditate over daily and sensible normal practices.

To discover the possible causes of additional costs, we created an instrument that did not lead any respondents. Also, to provide a degree of internal validation, we needed to use an approach that sampled a fairly large population to allow us to generalize the results and develop a reasonable assessment of where there was variability.

Because the self-defence process is so well ingrained in how we do business, identifying the process elements that add no value, and their associated cost, required care and a sufficiently large sample.

The original study that we undertook encompassed 250 respondents from across Canada. The sample included 50 owners, 45 consultants (designers) and 155 contractors. The sample included construction specialists working in both heavy and light industry, institutional construction, heavy civil construction, and commercial and residential construction. For analysis purposes, six respondents from residential construction were excluded, as many of the practices, as well as the size and complexity of projects, are different in that sector.

The study identified the things that lead to added cost in construction projects resulting from exculpatory clauses and their use. It also gave us an indication of the cost of using these clauses and the utility in their use. Both the cost and the perceived value of the clauses were surprising. The surprise was so strong that I asked another researcher to repeat the investigation. The second study showed virtually identical results, suggesting that there is a real issue that needs to be addressed.

The study results were so surprising, that we repeated the investigation with a broader sample and with modified questions. The results were almost identical.

6.1.4 Limitations
As with all studies, there are real limitations in their applicability, accuracy or significance. These limitations are the result of practical considerations in developing the study methodology and collecting the

information. Accuracy of the replies offered by the respondents is also hard to gauge, although the remarkable consistency lends credence to what we discovered in both studies.

Driven by the surprising results, I had the study repeated. This time we used a different approach. Also, we extended the study to professionals and experts in the United States. The results were very similar, suggesting that the findings were reasonably accurate and were applicable in North America, not just in Canada. Although we have not tested the concept beyond this geographical boundary, I have worked on projects and on contracts in many other countries. When there is a legal framework based on common law and when the contracting process has been influenced by Western thinking, there is a significant possibility that there is a cost (and arguably a significant one) associated with the use of exculpatory clauses.

 The initial study was conducted in Canada.

 The second, somewhat broader study suggests that the results are applicable at least in North America and probably anywhere where the judicial system is based on common law.

 We do not know whether this study's results are valid beyond Canada and the United States, but all indications point toward this as being likely.

6.1.5 Validation

Can we validate the results? The answer, at least in the North American context, is that there are several strong indicators that the results are meaningful. This is based on the following:

- Two separate studies with different sample populations produced similar results.
- There is a significant number of stories and cases that corroborate the findings of these studies.
- Reviews of the findings by industry experts appear to confirm, not challenge, these findings.
- Statistical analysis confirmed the significance of the results.

- The two separate studies were examined for consistency and were found to have respondents who could practically be considered to be from a similar population.

When we look at the highlights of the studies into the cost of exculpatory clauses, the message that emerges is clear. In today's competitive world, we can no longer afford to <u>over</u> protect ourselves against unscrupulous contractors. How we deal with this and turn the problem into an opportunity is described in Chapter 12. To fully understand the rationale behind the proposed approach to procurement and supply-chain management outlined in Chapter 12, we need to consider the issues raised in all of the preceding chapters. That said, we see two important findings emerge. First, we see that there is a significant cost associated with inappropriate risk assignment. Second, we see that there is little or no perceived value in assigning such risks – just the cost.

Boss, I think I've found the perfect guy
to audit our safety practices on site.

6.2 The Cost of Assigning Risk: What We Found

6.2.1 Cost of Weasel Clauses

Both studies demonstrated that there was a mean cost of between 8% and 20% that could be attributed to the existence of five of the most common exculpatory clauses. The two most significant factors that affected the range were market conditions and who the key people involved were. These factors changed the mean from the lower number in a buyer's market and with high trust to the higher one in a seller's market and with low trust between buyer and seller. If we assume that normal business cycles between these extremes, then we can take the average of these two extremes and round it off to the nearest 5% to reflect the probable accuracy of the studies. This means that, on average, we spend 15% of the price of construction on something other than construction. If the profit made by most general contractors is about 2% of turnover, then we can quickly see that this money is being used in some way rather than returning to the contractor as a part of its profit. So where does it go?

Depending on market conditions and trust levels, between 8% and 20% of the price of a construction contract directly relates to the premiums associated with response to the presence of exculpatory clauses. This translates to an average of about 15% of the money spent on construction being linked to these risk-assigning clauses.

Before we look at where the money goes, we should know what the clauses were. The following clauses were taken from a contract form that is the basis of a significant number of construction contracts in Canada. Similar clauses are used in other contracts in North America and elsewhere.

Clause 1: No Damage for Delay
". . . the Contractor shall not have any claim for compensation for damages against the Owner for any stoppage or delay from any cause whatsoever."

Clause 2: Examination of Work
"The bidder is required to investigate and satisfy himself/herself of everything and every condition affecting the Work to be performed and

labor and material to be provided, and it is mutually agreed that submission of tender shall be conclusive evidence that the bidder has made such an investigation."

Clause 3: Examination of Engineering Work
"Any representation in the tender documents were furnished merely for the general information of the bidder and were not in any way warranted or guaranteed by or on behalf of the Owner or the Owner's Consultants and its sub-consultant, employees, and neither the Owner and his/her Consultants or employees shall be liable for any representations, negligent or otherwise contained in the documents."

Clause 4: Liquidated Damages
"If the final date of completion of the Contract Works according to the agreed delivery time is delayed and if such delay is attributable to fault of the Contractor or its representatives, then the Contractor shall pay the liquidated damages, and not a penalty, the amount of one thousand dollars ($1,000) to Owner for each day that expires after the Contract Time specified in this Contract."

Clause 5: Indemnification
"The Contractor shall be liable to Owner for all losses, damages and expenses whatsoever which Owner may incur, and in addition be liable for and shall indemnify, and hold harmless the Owner, its officers, directors, employees, consultant and agents against and from all proceedings, claims, losses, damages and expenses whatsoever which may be brought against or incurred by the Owner including solicitor and own client (indemnity) costs; as a result of claims, demands, actions or proceedings made or taken against the Owner by persons not parties to this Contract. Such indemnification shall survive termination or completion of this Contract."

6.2.2 Value of Clauses
In addition to looking at the cost of clauses, we also set about assessing the value of the clauses. We considered the logic that contractors only have one source of income – their customers. If they are still in business, then their customers must – in some fashion – provide the money to cover the costs associated with the risks that they take in doing their work. If there was a cost, we argued, there must be a value. It turned out that there was no significant value perceived by the respondents.

Of the five clauses we used, the respondents commented as shown in Table 6-1.

Exculpatory clause	Clause NOT necessary	Project objectives NOT served	Clause a subject of disputes
No damage for delay	92.5%	98.1%	79.2%
Examination of work	67.3%	67.1%	84.2%
Examination of engineering work	93.4%	96.6%	80.4%
Liquidated damages	75.5%	75.8%	79.0%
Indemnification	83.5%	86.4%	85.0%
Overall average	**82.4%**	**84.8%**	**81.6%**

Table 6-1. Qualitative evaluation of exculpatory clauses by all parties. (From Khan, 1998).

Overall, more than 80% of the respondents felt that the clauses were unnecessary, did not serve the objectives of the project, and were likely to lead to disputes.

 The utility (value) of these clauses is effectively close to zero.

This begs the question: Why are we still using such clauses? There is a logic behind such a view. If we have a problem that is covered by one of these clauses, we are likely to have a dispute. If we resolve the dispute, we probably need not have had the clause in the first place. If we do not resolve the dispute, then we will have a claim from the contractor. If this is paid, then again, we have no value in the clause. If we take the claim to court and the contractor wins, we have lost – again, there is no value. If we "win" the lawsuit, then we need to check on the cost of such a suit. One study by a Toronto law firm suggests that the average $100,000 construction dispute costs each party more than $140,000, excluding expert witnesses, transcript fees and lost opportunity costs. Based on recent experience, it is likely the cost in the United States is higher than in Canada.

As the French writer Voltaire once said, "I have only been ruined twice in my life. The first time was when I lost a lawsuit; the second, when I won one!"

 We need to ask ourselves why we persist in including these clauses in our contracts.

6.2.3 Things that Affect the Premiums Charged

There are several factors that contribute to this additional cost to the buyer that is built into the price of construction today. Zainul Khan's study, a Master of Science thesis completed at the University of Calgary, helped us understand how this premium is spent.

The overall ranking of contributors to premiums based on all parties' assessments is shown in Table 6-2.

Factors contributing to increase or decrease of risk premiums	Weighted score*
Unforeseen site conditions	285
Technical complexity	267
Contract terms	262
Environmental risk	235
Degree of hazard in the work	215
Need for work	-172
Location	166
External factors	163
Project complexity, size, and duration	159
Economic conditions and market risk	154
Design completeness	108
Stakeholders concern	104
Owner's payment capability	97
Contracting parties' relationship	61
Contractor's expertise	-53

Table 6-2. Effects of factors on the premiums charged (From Khan, 1998).

* Negative numbers indicate a reduction in premium. The score is calculated by multiplying the frequency of the algebraic sum of respondents who identified the factors as adding to or reducing the overall premium for risks identified in the five clauses in the study by the mean assessed effects.

Table 6-2 lists factors that, in the opinion of the respondents, had a positive or negative influence on the premiums. Table 6-3 is perhaps a bit more interesting, as it tells us how much, on average, the respondents believed is allocated to the risk premium and what the value of the premium is as a percentage of the contract price.

Cost components	Risk premiums for each clause (% of total price)					Total risk premium for all clauses (% of total price)
	Clause 1	Clause 2	Clause 3	Clause 4	Clause 5	
Time-dependent job site costs	0.8	0.4	0.4	0.7	0.8	3.1
Contract administration costs	0.4	0.4	0.4	0.4	0.4	2.0
Choice of management team (pay)	0.3	0.3	0.3	0.3	0.3	1.5
Legal fees	0.5	0.4	0.5	0.5	0.7	2.6
External consultant and expert fees	0.1	0.4	0.5	0.1	0.2	1.3
Insurance, bonding, and other charges	0.5	0.4	0.4	0.3	0.7	2.3
Additional planning	0.3	0.3	0.4	0.3	0.5	1.8
Special equipment costs	0.2	0.2	0.2	0.4	0.2	1.2
Overtime allowances	0.3	0.2	0.2	0.4	0.2	1.3
Additional overhead cost	0.5	0.2	0.3	0.3	0.5	1.8
Total	**3.9**	**3.2**	**3.6**	**3.7**	**4.5**	**18.9**

Table 6-3. Contractors' risk premiums for adverse conditions (From Khan, 1998).

Almost 19% is the premium associated with these clauses in adverse conditions. Adverse conditions were defined as being those in which the contractor has a high need for work, the project is technically complex, the contract administration by the client is known to be unfair, the contract award is based on price and the contract design is incomplete.

Favorable conditions yielded a premium of about 9%. These conditions were the antithesis of the adverse ones enumerated in the preceding paragraph.

The second study, undertaken to challenge or validate the findings, by Zaghloul produced comparable premiums in the range of 8% to 20%.

Of note, the cost factors that generated the premiums add nothing to the end value of the project. They are simply the cost of doing business in a confrontational environment.

In addition, the premium used (total value of 19% based on the Khan study) considers the more adverse conditions contemplated in this study.

6.2.4 Ingrained Habits and Mistrust

Our habits, when it comes to the procurement process, are deeply ingrained. They are the result of years – probably generations – of mistrust. There is so deep-rooted a perception that we can scarcely conceive the notion that a vendor will actually volunteer to help us as a buyer. Not only that, but we are schizophrenic about this; as a buyer, we are careful not to trust a vendor. As a seller (and most buyers are sellers too), we like to build a relationship with our customer so that they can trust us. Through this trust in us and our product or service, we build loyalty and repeat business. Our habit of mistrust as a buyer is expensive. We have just seen that.

In the buyer's habits that we see and that the findings of current research confirm, we try to eliminate "soft" issues such as trust and relationships from the procurement process. The bigger the ticket item we are buying, the more likely we are to try to make a fully auditable and "objective" decision. A bad decision made for the apparent right reasons is invariably better (and safer) than a good decision based on a gut feeling or some other cosmetically unacceptable reason.

Even today, it is depressingly common to find contract administrators in sophisticated buyer corporations who believe that we need to eliminate trust from contracts.

In larger organizations, the buying and the selling functions tend to be separated. It is fairly unusual for a buyer in an organization to also be or have been a salesperson. This means that the disparate perceptions or expectations of the people in the two functions often do not get looked at internally in the organization. They are merely looked at when we come into contact with another party trying to sell us something or when we are trying to sell something to someone else. It follows that the expectations of buyers and sellers, as well as their perceptions of what is happening, do not always overlap. One of the important points that relates to this is the understanding by buyers of the premiums we have just looked at. Our most recent study (Zaghloul, unpublished) suggests that the vast majority of contractors are aware of – and can quantify – the premiums they charge to cover risk. Equally, it confirms that most buyers are blissfully unaware of them.

There are many misconceptions and much misunderstanding around the issue of trust in a business deal involving exchange of goods and services for payment.

One view of the rationale behind the use of clauses such as the ones listed in Section 6.2.1 may lie in a different, although not incompatible, interpretation of the world of contracting. Increasingly, people are accepting that contracts are really just a basis (or set of rules) for subsequent negotiation. Once a buyer has awarded a contract, the feeling is that there is little power left to deal with the inevitable changes. In the absence of power, buyers feel exposed and vulnerable to being taken advantage of by the contractor or supplier. Positioning for maximum power may well be a reason why current practices in the use of exculpatory and other risk-assigning clauses persist.

Much of what we see in human behavior around the issue of trust may have to do with power and negotiation.

Given the practical limits of implementing change in the practices associated with how we buy goods and services, there still seems to be some opportunity to improve on the final price of contracted work by reducing the premium associated with inappropriate assignment of risk.

There are barriers. First, we need to have a will to try to make a change. This needs to come from the buyer organization. Historically, seller-driven initiatives to modify or improve contracting practices have failed. Then we need to persuade the lawyers and other advisors that there may be some benefit in not doing something that only appears to increase risks and, thus, cost money. Generally, we have asked lawyers to protect us. We have not asked them to consider or identify the cost associated with such protection. Once we are past this point, we can start to eliminate the exculpatory clauses. Early evidence would suggest that this alone will not reduce the premiums, as many of them are deep-rooted in the culture and habits of all of the parties. The work to find the best way to recover the lost cost inherent in our current practices continues, but there are some indicators of how this can be done in the evidence that has been collected to date.

At least part of the cost associated with inappropriate risk assignment and the associated business practices is avoidable. The challenge lies in how we may eliminate these risks.

6.3 Eliminating Risk Premiums

It seems logical that the money spent on projects and that delivers no real net value need not be spent. This simple thought led to the idea that the risk premiums associated with exculpatory clauses and the associated positioning, power plays, negotiation, risk management and other behaviors could be reduced if not eliminated entirely. Practitioners have done several things to address some of these issues.

A number of approaches have been tried to achieve this objective, but all have resulted in only very limited success. In the North Sea, the Cost Reduction Initiative for the New Era (CRINE) has produced some interesting outcomes, based on the principles that are now embedded in many partnering agreements. Other initiatives have also led to cost reductions, although fewer of them have the profile of CRINE. A study by the Construction Industry Institute (CII) demonstrated the correlation between increased trust and cost savings. The Calgary Method for more effective risk allocation came out of work I did to earn my Ph.D. In summary, what are these initiatives?

1. CRINE focused on more effective supply-chain management through risk sharing and making success or failure common to both buyer and vendor.
2. The CII study demonstrated that, within sensible and practical limits, increased trust between buyer and vendor leads to reduced cost.
3. The Calgary Method helps to identify risk premiums on real projects and then compare bids based on more intelligent assignment of risk to buyer or seller.

Although each of these initiatives appeals to the logical side of professionals in the business, they have two significant drawbacks. First, they are new ideas, so they inherently represent a change in an industry that is reluctant to adopt such change. Second, the change is one that is apparently counterintuitive to anyone who perceives contracts and contract administration as a defensive process, protecting their employer or host organization from the ravages of their suppliers and contractors.

ALL of the drawings on this project are blank.
We signed a non-disclosure agreement with the client.

My father once said that half the solution to a problem is knowing that you have a problem in the first place. This is consistent with the view of many practitioners that Zaghloul polled in search of a way to change the industry. The first piece of advice we got was to educate. As to who needs the education, the consensus was that, although everyone needs heightened awareness of the issues at stake, the owner or buyer is the one who can effectively instigate the necessary changes. Some of the insights presented stem from early results of the as yet unpublished Zaghloul study.

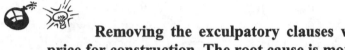 **Removing the exculpatory clauses will not reduce the price for construction. The root cause is more deeply ingrained in the process.**

6.3.1 Awareness of these Costs

Education is a key factor in addressing the issue of risk premiums that do not really need to be there in the first place. Lack of awareness of the problem, blame cultures in buyer and seller organizations, and the need for self-preservation all lead us to keep doing what we have always done. It is easier, there has previously been no evidence that there is a cost associated with assigning risk to others, and the result is that we still get competitive prices if we bid the work. The fact that all of the prices from all of the bidders are inflated masks the opportunity to do better through more astute risk management through contracts.

 If we want to solve a problem, we first need to acknowledge that we have a problem.

Both an intuitive assessment of what is going on and the research that we conducted twice to be sure we had something real tell us that giving someone else a risk will cost us something. So what can we do about it? And, before we proceed, how do we know that our solution may work?

The evidence that there are significant costs associated with exculpatory clauses is irrefutable. The open question relates to how we can eliminate these costs.

The real incentive to do something lies in the competitive advantage that a significantly lower capital cost for infrastructure and other investments will have on the business we operate. If we can reduce costs with little or no real added exposure, this is a natural step to take.

The first companies to reduce the cost associated with exculpatory clauses and to eliminate the money that is wasted in misallocation of risk will gain a significant commercial advantage through the reduced cost of doing business.

6.3.2 Awareness of the Issues

If we look at the cost-contributing items in Table 6-3, we can see that they relate to process ('social' issues) more than anything else. These items are reproduced in Table 6-4.

Cost Components	Premium (% of bid)	Nature of underlying issue		
		Technical	Business	Social
Time-dependent job site costs	3.1	X		X
Contract administration costs	2.0		X	X
Choice of management team (pay)	1.5			X
Legal fees	2.6			X
External consultant and expert fees	1.3	X		X
Insurance, bonding, and other charges	2.3			X
Additional planning	1.8	X	X	X
Special equipment costs	1.2	X	X	
Overtime allowances	1.3		X	X
Additional overhead cost	1.8	X	X	X
Total	18.9			

Table 6-4. Nature of cost components.

Table 6-4 may look wrong at first glance. I am trying to illustrate the UNDERLYING cause of a cost overrun. An example of a technical issue would include addressing a specific requirement of the contract or the specifications. An example of a purely objective business issue would include the cost of special equipment or the contingency reserve for overtime. An example of a social issue would include a gut feeling, a sense of uncertainty, a need to feel more protected, morale or motivation issues, a decision of whom to work with, or some other such decision. Error and omission insurance can be used to illustrate the thinking behind determining the nature of the underlying issue.

Why do buyers of professional services often write into their agreement to purchase these services a condition that requires the architect, engineer, or other professional to carry a specified amount of insurance coverage for errors and omissions? Does this create a sense of concern that will affect the design? Who really pays for this (see Overkill in Section 2.3.3)?

The human condition is such that we tend to act in our own interests. If we raise the specter of risk, liability, or other exposure, we tend to generate protective behaviors that mitigate against the best technical or business solution for all concerned. A perceived threat of being sued if the design is wrong will almost inevitably lead to ultraconservative designs. No innovation. No creativity. Little chance that there will be any breakthrough performance in the project.

There are technical, business and behavioral issues that need to be addressed. The technical ones are the easiest and the behavioral ones the hardest. Unfortunately, the latter have the potential to yield the biggest savings.

Perhaps the simplest part of taking the cost of risk out of the price of a contract is to address the issues that are purely mechanical. These do not involve a business risk or any decisions by legal counsel, management or others who wish to provide protection at any cost to the organization or even to themselves. The purely mechanical aspects of managing risk are relatively easy to quantify (e.g., insurance coverage and who provides it, allowance for bad weather, planned overtime). The reason for this is simple. The decision is based on a definable need and an appropriate response. The assessment of the risk is relatively straightforward.

The technical issues revolve around the mechanics of risk allocation and subsequent management.

Business decisions regarding risk, who should carry it, what the premium is, and whether it is worth paying it or covering the risk in some other way are hard processes, as they require a degree of judgment. This, in turn, exposes the person making the judgment to criticism. Business decisions are affected by the culture of the organization, including its propensity to take risk, the extent to which a blame environment exists, and other less tangible factors. Business processes and practices play a significant role in the formulation of risk management decisions. The history of a company and the individuals in it are factors that will affect the judgment of the people who make the decision affecting risk allocation through contracts. If we feel exposed, we will be more conservative in assigning risk. Rarely, it seems, do buyers consider what it may be costing their organizations when they pass risk onto a contractor or supplier. The normal process of assembling projects and teams also helps in obscuring the real cost. For example, what is the outcome if a buyer asks a design consultant to produce a design and holds the designer liable for performance of the end product? In addition, if that liability includes any consequential damages, then how will the designer react? If they are good business people, they will pass as much of that risk onto the contractors and suppliers who are needed for the next stage in delivery of the project. This is done by making sure that the specifications and other parts of the contract have the necessary clauses to reassign the risk. There is no cost to the designer in doing this. There is a premium that will effectively be hidden from the ultimate buyer by the process, especially if there is a requirement for all bidders to be compliant if their bid is to be considered.

Now, if we are the contractor, will we pass much of the risk onto our subcontractors and suppliers? My personal experience as a contractor and as a subcontractor is that we would not hesitate to pass on the risk. We do not even know the full premium associated with the original decision of the owner to pass on risk. Nor does the designer. And the owner remains blissfully happy as the premium is now completely hidden. The results of Zaghloul's work mentioned earlier in this chapter are totally consistent with this set of observations.

The business risks relate to business processes and practices. A more appropriate balance between the cost of managing a risk and the utility gained from the selected risk management approach will yield cost savings.

Much of what these studies have found points at what may be interpreted as irrational behavior. Some of the buyers who know there is a premium also know that there is little value added to the project or the process by use of exculpatory clauses, yet they still use them. Reasons offered for this apparently strange behavior are usually founded on corporate policy, reticence of the legal department to consider change and even industry practice. It could be argued that this makes sense in the absence of trust. But when trust exists, then the rationale becomes arguably weak. Does this mean that we can connect mistrust to the behaviors we see in the use of exculpatory clauses?

Hi, you must be the architect who forgot to pay your professional liability insurance premium.

In the absence of a high level of explainable and auditable trust (does this exist?), we are left in a position in which current behaviors and the perception of what is sound business practices will likely remain in place. The obvious element that we must therefore change would appear to be the level of trust that we may enjoy between contracting parties.

 The behavioral issues revolve around trust. The problem with trust is that it is ethereal and so hard to measure and audit – at least until it is too late.

6.3.3 Trust and its Limitations

Blind trust was identified as the point at which there is no justification for changing our behaviors. This was specifically identified in the CII study of the relationship between cost and trust on construction projects. Even before we get to the point at which we blindly trust, we may be exposed, and this is a concern. However, in the absence of any trust, we add a significant overhead to the cost of doing business. When is the right balance for optimal trust levels commensurate with good business practice? There cannot be a specific answer, so we must look for indicators that can guide us in each situation we encounter.

 If trust is the basis on which we can expect the best potential return, we should understand the basis on which such trust may be used with minimal risk or exposure.

There is probably a practical limit to the extent to which we should reasonably trust another party.

One option is to look at the use of contracts. If both parties have a stake in the successful outcome of a project, then work can – and often does – proceed without a formal contract in place. If we work in this way, then the ground rules are fundamentally different. We do not have a formalized set of rules for our behavior toward the other party. Under these conditions, if there is a prospect of future reliance on the other party for the success of our business, then we have a potential way of working together in which we are more likely to collaborate and help each other than we may otherwise do under a formal and confrontational contract.

Relying on trust rather than on a watertight contract fundamentally changes the rules.

Whenever two parties work together, an implicit contract exists. The terms of the contract are hard to determine, though, in the absence of a formal document that articulates them. The principle is referred to as *quantum meruit*. This is Latin for what may be translated as "as much as is deserved". The principle is simple and sometimes articulated in court opinions. The plaintiff in such a dispute may recover the reasonable cost of providing goods and services that may have been enjoyed by the defendant and under circumstances in which the plaintiff may reasonably have been expecting to be paid for such provision.

Under the right conditions, trust can be legally enforceable.

However, we generally do not want to go to court every time we use such an approach. Taken on its own, relying solely on trust means that we rely on not having to sue to recover our costs or other entitlement as a result of a formal or informal (such as verbal) agreement. This requires the good will of both parties to the contract. This model is similar to what we have doubtless heard of – or even used – as the "handshake deal" which apparently was more common in the "good old days" than it is today. Perhaps this is a reflection of the global economy, of more mobile people, and of the reduced commitment to preserve our reputation and relationships in our own community if we want to stay in business.

On its own, trust relies on the ethics and good will of the parties involved.

A well-formulated contract provides a set of rules for many of the actions and decisions that follow. The contract is what we measure performance against. It allows us to audit what happens.

 One significant drawback of purely trust-based relationships is that they are difficult to audit without a paper trail. Such a trail, if done rigorously, appears anathema to trust.

6.3.4 Auditability

Most people who work for any organization other than closely held corporations or small family businesses routinely face the need for full auditability. We have, in other words, got to be able to demonstrate to shareholders or taxpayers or other critical stakeholders that we have behaved and acted in the best interests of the organization we are working for. This need will not go away. The safest way of demonstrating our good intentions has actually got nothing to do with looking after the organization's best interests, but everything to do with looking after our own. It is safer to award a contract to the lowest bidder, even if we suspect that this bidder will cost us more overall, than to award it to any of the other bidders. It is hard to explain at the time of award why we would do anything that costs the buyer any more than the lowest price. This concept was discussed in Chapter 4. The same issue arises over changes, dispute resolution, progress payments and quality assurance as we manage the contract.

 One of the reasons we develop contracts is that, under normal conditions, the contracting process provides a reasonable audit trail.

6.3.5 Value for Money

If we are truly interested in obtaining the best value for our money, then the traditional approach to bidding and awarding work is probably not the most effective. We need to look at other options for selection of suppliers of goods and services. We also need to look at what we include or exclude, state or imply in the wording of our contracts.

There is a significant cost associated with how we do business based on low trust that has the potential to be eliminated. Eliminating this cost must produce a product that offers greater value for the reduced cost, simply because it is the same product and it cost less. If, in developing better working relationships between the contracting parties, we end up with better communication and more creativity, then we are likely to end up with a better product, not just a lower cost and faster delivery. Now THAT is real value for money.

 At the end of the day, any process that reduces the total cost of contracting out goods or services leads to a net cost savings.

If we are really concerned over the risks and wish to stick with current practices, this is a legitimate position to take. If, however, we are willing to try something new or different in the hope of getting a better result, we can have some confidence in the potential for success. This is demonstrated by a growing number of initiatives, such as CRINE and its Norwegian equivalent in the North Sea and individual projects and cases that are reported from time to time in journals and trade magazines.

 If there is a risk associated with a cost savings, the net savings may still be legitimate if the exposure on the inherent risk is less than the expected savings.

 Chapter 6 illustrates that there is a cost associated with exculpatory clauses. The value of these clauses is debatable.

Chapter 7 addresses the role of trust in contracting.

Chapter 7

THOU SHALT TRUST THY CONTRACTING PARTNER, BUT NOT DO SO UNREASONABLY

> *The role of trust in contracting: types of trust and trust profiles.*

Until we experience it for ourselves, the idea of tying project outcomes to trust seems strange for most practitioners in the business of delivering projects that rely on third parties or on contractors. Yet in 1993, the Construction Industry Institute (CII) based in Texas, undertook a visionary project to examine just this phenomenon. Since then, the interest in the role of trust in business has grown significantly.

Part of that growth has been in the form of other research; one of the larger studies in this area was started at the University of Calgary in 1996. This multimillion-dollar investigation has already yielded some interesting findings, some of which are presented in Chapter 7, in conjunction with work done by others.

One of the fundamental discoveries of the Calgary University work has been the classification of trust types for the purposes of understanding trust in a business context. It would appear that there are three types of trust. Chapter 7 describes these in more detail.

Building on the work of CII, to further understand the relationship between cost and trust, Chapter 7 addresses the role of trust in the contracting process.

7.1 The Cost-Trust Relationship

In 1993, CII presented their findings from a study of the relationship between cost and trust on construction projects. The work was presented at special conferences run by CII. It has been published in CII publications, the most easily read one being the "Cost-Trust Relationship", a document prepared for the 1993 Contracting Phase II Taskforce Breakout session at the 1993 CII Conference in Austin, Texas.

7.1.1 Cost Elements

To determine the likelihood of a cost being associated with a certain level of trust, CII set about defining the types of cost that may be affected, as well as providing a definition of trust itself. There were two types of cost that were identified. One was associated with higher costs resulting from the business steps that were added as a consequence of mistrust. The other was linked to cost-savings opportunities created as a result of high levels of trust. The first type of cost was labelled "transactional added cost". The use of the term "transactional" is important, as it clearly identifies the added cost as being associated with transactions and NOT with any added value. The second group of costs are a bit more complicated, as the reason for achieving the cost savings lies in more than one area.

The Construction Industry Institute cost-trust study identified and measured two classes of costs: transactional added cost and avoided cost.

This claim is for $200. I reckon the one-page request for $2 Million more will get through easily if we deal with this one first.

A natural problem with assessing the real value associated with elimination of added transactional costs lies in the fact that these costs cannot be accurately determined, as the transactions are not tracked on time sheets or in other ways that can lead to a potentially objective and fully defendable result. This notwithstanding, the process of determining the value, as used by the CII research team, had its own value. The standard approach to this dilemma of measuring the immeasurable is to rely on industry experts. Using sampling techniques designed to achieve objectivity can help eliminate most or all of the potential bias of the selected experts.

With a reasonable approach to measuring something for which hard metrics simply do not exist, the study team was able to distinguish levels of added or saved costs. The range went from extreme added cost to negligible added or avoided (saved) costs to extreme cost savings (or avoided costs).

It is interesting to note the use of the term "avoided cost" as this implies we spend money we do not need to spend in the process of contracting. This is consistent with observations from other sources, as well as being in line with the observations we can make from the examples sprinkled throughout this book.

The levels of cost ranged from extreme added cost to negligible added or avoided cost to extreme avoided cost.

The study was able to articulate the factors that contributed to the cost of projects. These factors are listed in Table 7-1.

Cost Factor	Respondents Who Identified the Factor
Project team efficiency	70%
Timing of decisions	67%
Project schedule	66%
Project performance/quality	58%
Timing of approvals	54%
Amount of rework	47%
Administrative costs	45%
Field supervision	42%

Cost Factor	Respondents Who Identified the Factor
Completeness of project scope	39%
Project safety	34%
Materials/equipment deliveries	31%
Contingencies	24%
Legal costs	22%
Interest costs	14%
Overhead multiplier	10%
Cost of bonds	7%
Insurance costs	5%
Other	3%

**Table 7-1. Ranked cost factors
(Construction Industry Institute, 1993).**

The data in Table 7-1 demonstrate that there are a number of factors that most stakeholders were aware of and could assess as contributing to the added cost of a project in the absence of trust. It is interesting that the first item addresses team efficiency – a factor that is based on interaction between people –and the second one has a similar basis. The other observation of interest is that this study differentiated the people who made decisions (timing of decisions) and who approved things (timing of approvals) from the "team". Today, the most effective project teams are much more inclusive.

It is also interesting to note that there are no cost factors that directly contribute to added value for the buyer. There are only a few that could arguably contribute to value – and then only indirectly.

Cost factors were identified and ranked by the percentage of respondents identifying the factor.

The costs that were identified are quite consistent and the fact that a significant number of respondents identified them suggests that they constitute a reasonably reliable list. The absence of objective measures of these costs reinforces the increasing awareness that we tend to measure the wrong stuff when we manage businesses and the projects that drive them. This means that the listed costs were assessed subjectively by the respondents. This is not necessarily a bad thing, as we know that there are practical limits to the effective tracking of costs using current accounting principles. How do costs get allocated to cost codes anyhow? Are people likely to freely and completely honestly allocate costs to cost codes that identify the expenditure as non-productive or – worse – as downright wasteful of corporate resources?

The costs were assessed subjectively by the respondents because avoided costs are never recorded and the added costs were not considered accurate.

7.1.2 Trust Components

The study defined trust. This is a common step in most research. The argument behind this is that if we cannot define something, how can we research it? In the work at the University of Calgary, we have tried hard to not define trust, as any definition is limiting and may stop us from making a critical discovery. So far, it has not proven to be an obstacle in our work.

This said, the definition used in the CII study is interesting, as it describes two types of trust that are wholly consistent with the trust model that is used in the Calgary study. The first type of trust is competence, and the second is integrity. In the CII study, these manifest themselves in the following definition – taken from the CII study summary report (1994).

"The confidence and reliance one party has in the professional competence and integrity of the other party to successfully execute a project in the spirit of open communication and fairness."

Note in this definition the use of the words "competence" and "integrity". These are distinct types of trust independently identified in the Calgary study.

The CII study identified, but did not separate, two important aspects of trust.

The CII study went on to identify three conditions that need to exist for trust to be present. The first of these is openness. Trust simply does not form and crystallize in the absence of effective open communication. Can we trust someone who appears not to trust us because they are holding back on something?

A lack of flexibility is manifest in certain types of behavior. A behavior we associate with inflexibility is insistence on compliance with an agreement even if it is patently the wrong thing to do. We expect others to follow a bureaucratic approach to delivery of a product or service or we demand absolute compliance with specifications even when something better could be done by non-compliance. And there are many more examples. We get the drift. The point is that there is usually a similar action that, when we are responsible, we see in a different light. We would see the issue of compliance as getting what we had contracted for, not something that may be different or less. Inflexibility readily turns into assertiveness and protection of our rights.

The issue of fairness is equally one of perspective. As long as expectations are out of line, we can safely assume that the perception of our fairness is that we are more than fair and that the other party's fairness is the inverse.

Our view of reality often depends more on perceptions than on reality itself – especially when we are dealing with issues such as openness, flexibility and fairness.

The study identified and used three trust conditions: openness, flexibility and fairness.

One of the most important findings of this study lies in the pragmatic value derived from assessing trust levels. In the end, we deal with people. We may work for a company or an organization, but it is our views that are reflected in how we conduct business on behalf of our organization. As people, we respond well to being communicated with. We prefer to work with people who we know are going to be honest. We are generally happier if we know what is required of us and if we can communicate our expectations to the people we need to work with.

A significant conclusion of the study was that assessment of trust levels between contracting parties is a potentially useful part of the contracting process.

There were eight trust-related issues identified in the analysis of the information garnered by the study. They were:

- Open discussion of alternative methods of performing the work
- Value engineering
- Constructability
- Contract administration
- Risk allocation
- Level at which disputes related to risk allocation were solved
- Communications
- Dispute resolution.

Once again, the study came up with an interesting list that has since proven to be of significance. The cost associated with risk allocation we now know to be high. Raising the level at which disputes are resolved also adds cost. This was demonstrated by CII in their study "Dispute Prevention and Resolution" (1993), which showed that the higher the level, the lower the chance of coming to a quick and cost-effective resolution. Other work on constructability and value engineering repeatedly identifies the absence of trust as a barrier, or the need for trust to be effective, in delivering good results.

 The study identified eight trust-related issues, each of which is significant and contributes to improved procurement efficiency if properly addressed.

7.1.3 The Relationship between Cost and Trust

The cost-trust curve produced by the study team shows that there is a clear relationship between trust and cost. The higher the trust levels, as measured by the study, the lower the ultimate cost of the project. Simple really.

As trust increases, so the probable final cost of the project decreases.

Even the most case-hardened practitioner would have problems resisting the reality of this conclusion. Even the greatest research cynic –

me included – would see *some* possible value in the results. The methods may be challenged. The value of the empirical data and the potential for error exist, but the message that there is a relationship between cost and trust is still very clearly there.

Independent study and observation have supported this conclusion. The work done on real projects in the North Sea under CRINE (Cost Reduction Initiative for the New Era) supports the finding that there is value to be extracted from the procurement process through people working more closely. Clearly, closer collaboration is linked to increased trust.

Even with the limitations inherent in this type of study, we can see that there is something of real importance in the concept of trust.

7.2   Trust Types and what they Affect in the Contract Relationship

Now that we have seen that there is a relationship between trust and improved performance in the business of contracting, it may be of use to look at trust a bit more closely. Before we do that, though, it is probably worth understanding the rationale behind the view of trust we are about to look at. This view is the result of several years of study and investigation that came about because of the findings of the preceding five years of research into better ways of managing projects and businesses driven by projects. We saw trust as a key ingredient of success in ALL of the previous studies, whether they address more effective teams, modelling, cost-reduction processes, value engineering, business over geographically distributed centers, or virtually any other aspect of effective management of projects and their teams and supply chains. Given this, and the enormous body of knowledge in the trust arena, we looked at the issue differently. First, trust WAS recurring as an issue. Second, trust research in a business arena was just getting going. Third, most of the existing trust studies had been conducted based on very specific definitions of trust. Many of these definitions were exclusive and were not transferable to other situations. For example, the CII definition cited in Section 7.1.3 hardly works in a father-daughter relationship!

So we decided not to look for a definition of trust. Furthermore, we decided that we would not try to understand the psychological significance, as this did not seem to matter (at least superficially) to the average business person. What mattered was the MECHANICS of trust –

so this is what we focused on. What emerged was a framework for understanding trust.

Trust means something different to everyone. The Color Trust Model was developed to overcome this challenge, and to focus on the business mechanics of how we develop, maintain or lose trust.

In the Color Trust Model, we took a specific approach designed to help us understand the way we built trust in a business context. Because we wanted the model to be independent of industry, culture, role or any other significant variable, we built it based on business needs first, then we tested it in a growing number of situations that eventually encompassed non-business situations. The model evolved and now has three distinct types of trust. Each type of trust behaves differently and is assigned a color. The three (pigment-based) primary colors are used. The three types of trust are listed here.

- Competence (Can you do the work well?). This is BLUE trust.
- Integrity (Do I trust you to consistently look after my interests?) This is YELLOW (or golden) trust.
- Intuitive (Does this feel right?) This is RED trust.

There are three distinct types of trust.

Both competence and integrity are identified in the definition of trust used by the CII study. The full model is complex and will, I hope, one day be the subject of a whole book. For Chapter 7, we use the three types of trust in their simplest form. As such, they should help us understand some of the basic needs of better contracting using trust as part of our "toolkit". The explanation of the model may fall short of being complete because Chapter 7 focuses on the key issues as they relate to better contracting.

An earlier version of the Color Trust Model is explained in a bit more detail in the book "Don't Park Your Brain Outside" (Hartman, 2000).

The Color Trust Model identifies three types of trust. Two of these are implied in the work done by others, including the Construction Industry Institute. The third is new and, it turns out, both complex and critical to business management.

7.2.1 The Color Trust Model

The "Blue Chip" competence type of trust is auditable and addresses the level of confidence we have in another party to do specific work. This type of trust can be built based on track record, references and reputation. It is, therefore, transferable. We rely on this type of trust when performance is an issue to us. We do not like to retain professionals, be they architects, brain surgeons, lawyers, or engineers, if we have doubts about their ability to do the task that we have retained them to do. Most professions are regulated in some way, so we can normally rely on a minimum level of confidence in their competence if they are active and current members of the appropriate professional association. Beyond that, we may rely on reputation. Or we may ask for references from past clients. Alternatively, we look for evidence in past projects or performance. This type of reputation takes years to build and tends to be stable.

  Blue trust represents the answer to the question, "Can you do the work?"

Integrity trust is less stable than competence trust. This "golden" type of trust normally does not survive being damaged or broken. Integrity trust is built in part only by reference or referral. We need to confirm our trust in others' willingness to take care of our interests on a much more personal and direct level. We do this to validate any referral or reputation. This type of trust occurs primarily between people rather than with an organization or institution, although the latter does carry some value. Yellow trust takes time to develop. It takes virtually no time to destroy. We need to keep demonstrating our integrity and trustworthiness to maintain the level of trust we have earned.

Yellow trust represents the answer to the question, "Will you consistently protect my interests?"

The most volatile type of trust is intuitive trust. It is also the most complex, as it includes both emotional responses and "rapid processing". The former we recognize as being our response to something that we simply have a personal reaction to. The latter may appear that way, but it is unconscious thought that happens at about 2,000 times the speed of conscious logical thinking.

We can recognize the latter easily enough through everyday examples. We may see a project schedule and our first reaction may be that there is something wrong. We do not know what it is that creates that reaction until we start looking for the discrepancy that triggered the reaction. The same happens when we meet someone and decide if they will fit in our team. Or we read a contract document and something makes us nervous. What is happening? We are making connections, drawing on knowledge and experience, and the pattern is one that we have a response to at the subconscious level. The greater our knowledge and experience, the more often and accurately we get these "gut feelings" and the more often they prove to be right.

The pure emotional response and the "rapid processing" response are often hard to tell apart. The response, though, is important, because if something does not feel right, we become nervous and then defensive. It is much harder (with good reason) to trust others when things do not feel right.

Red trust represents the answer to the complex question, "Does this feel right?"

There are two parts to the red trust question. One addresses rapid processing of thought, knowledge and experience. The other addresses our emotional response.

Now that we are aware of the difference between the two parts to red trust, we can pause and reflect on what may be causing us to feel comfortable or not with a particular situation. If we can differentiate between the emotional and rapid-processing parts of our reaction, we can assess the situation more objectively and then make a better decision based on a more rational response to both the facts and an understanding of our response to them.

This is how we explain the request for additional payment
to the Client, so he will feel foolish asking us for facts.

 If we can identify and manage the difference between rapid processing and pure emotion, we are likely to make more rational – and arguably better – decisions.

The description of the Color Trust Model in this book is both brief and superficial. The model is still under development and will be published in more detail once it has been more completely validated and tested. I hope that the basic idea presented here helps in dealing with the concepts and suggestions that follow.

 This description of the Color Trust Model in this book is brief and potentially dangerous.

7.2.2 Myth and Reality: Post-rationalization and the Need for Auditability

In most situations in which we have to make purchasing decisions on behalf of others (such as employers or clients), we need to be able to justify the decisions we have made. There are expected behaviors associated with the process of spending someone else's money. The process needs to be auditable. Other people need to be able to follow the impeccable logic and fairness of the decision we have made. There is no place for a "gut feeling," even though responding to such a reaction may lead to a better decision in the long term for our client. Fundamentally, it is risky to us personally to make the best decision if we cannot demonstrate at the time – or afterwards – that we have acted in the best interests of the client. In the end, the client pays for this in the form of premiums, the outcome of poor – but justifiable – decisions and other forms of inefficiency that result from the protective processes we have built in to the business.

In most situations, as described in Chapter 6, we need to make auditable decisions when spending others' money on contracts. Auditable decisions and right ones are not always the same!

We are regularly faced with the choice between doing the right thing and doing things right. The right thing in a particular bid assessment may be to discard the low bidder. We would do this because we suspect that it has made a big mistake but cannot find it, or because we know that the contractor we inherit from the decision will cost us a lot more in the end through claims, slow delivery, substandard performance or other factors. Discarding this bidder may not be doing things the right way. How should we respond to this situation? If we are spending our own money or that of our organization, we may be able to justify our decision even if it goes against the standard approach. Chances are, though, that if it is hard to justify, even though we know we end up with a substandard contractor, we will make the decision to go with the cheapest initial cost rather than the expected lowest final cost.

In the context of trust between contracting parties, the need for auditable behaviors can potentially get in the way of doing the right thing because we tend to prefer to do things right.

Often, when faced with the situation just described, we lean toward the personally safe solution of selecting the wrong contractor. The word "wrong" here is used in the context that the right contractor is the one that our "gut feeling" tells us we should pick. But life is short, and the client will complain – and we will be exposed – if anything were to go wrong with the decision we made on an intuitive basis. We cannot explain the real reason we spent an extra few dollars of the client's money to save a potentially much higher amount later. If we make the intuitive decision, we will never be able to prove that the other contractor would have done any better.

Red trust provides us with some powerful capability that is blocked by many of today's "acceptable" practices.

One Proactive Dollar is Worth 1,000 Reactive Ones

This project was a multibillion dollar one that was being built at the same time as a number of similar ones. This meant that materials, contractors, specialty trades people and other key resources were going to be scarce. I was asked to review the project and identify the risks. My team did this and produced a report recommending spending slightly more than $10 million to improve management of the supply chain. Spending this money in the marketplace that the project was facing would significantly improve the probability of timely completion and could help eliminate potentially significant cost overruns. Conversely, we suggested, NOT spending this money would lead to a potential delay of about a year and a cost overrun, excluding lost revenues, of about $1 billion. Our client read the report and said we were too pessimistic. Not only that, but it was very difficult to explain an expenditure of a further $10 million – especially if we saw no net cost reduction or schedule improvement as a result. When the project was finished, it was more than $1 billion over budget and was slightly more than 1 year late. Most of the cause-and-effect events that we identified in our report had happened. Although the extra $10 million may have been hard to explain, the $1 billion overrun was easy to spend because there was no choice in the crisis that resulted.

Spending money in a crisis is so easy because there often appears to be no choice. Normally, this type of crisis spending is to protect the investment made to this point in time. The crisis is its own justification in most cases because it is tangible and there are no options at this point. Making proactive decisions to AVOID possible future risks is not seen as insurance in all cases.

To harness trust to the full, we need to be able to explain some of the things that are hard to measure and that we therefore do not have records or metrics for.

7.2.3 The Effect of Auditability on Contracts

We <u>are</u> human! Self-preservation is in our nature. Given that we may have a family to feed (or, perhaps, a sports car to maintain), we are likely to want to protect our jobs and, through this, our income. Most of us are subject to some sort of audit in the role we fill of contracting out work on behalf of our organizations. If so, we err on the side of safety so that the audit and the auditors do not have cause to give us a hard time. We are safer if we follow company policy – even if this is a clearly more expensive option than making a hard-to-explain decision based on our knowledge and experience. I do not know of anyone who was fired for following company policy!

If, therefore, we want to improve on company policy or practices, we need to be in a position in which it is acceptable to take a risk of some sort. This risk should not be one that limits our career. It needs to be such that there is a real opportunity to improve performance, and it needs some formal sanction from the host organization. These two elements allow us to safely try new ideas. If our organization does not offer these safety features, our organization deserves the outcome of not changing, and it will likely eventually go the same way as dinosaurs, buggy whips and typewriters.

Self-preservation is important to all of us. This means that any improvement to current practices needs to address this key need of most individuals to be practical.

The interests of an organization need to be maintained, so we have a duty to the shareholders (or taxpayers in the public sector) to provide a degree of protection. If we accept that this should not be at all costs, then the cost of the protection should not exceed the risk we are addressing.

Now that we have seen the significant cost of exculpatory clauses and the inappropriate allocation of risk, we can look at this issue in a different way. We could argue that we may no longer be protecting the interests of the organization if we allow this cost to continue to take a bite out of the value we are trying to build through the work being done under such contracts.

Protecting corporate interests is important to an organization's stakeholders. We need to ensure that new practices reflect the ability to manage this.

In truth, we really have not been able to measure, in the form of hard accounting numbers, the real cost of these clauses. So we still have the excuse that the cost has not been absolutely proven. This excuse will remain until someone who buys and sells goods and services tracks the real costs in a meaningful way.

Knowledge of the real cost of not trusting still eludes us – despite the available, strong and consistent findings of research – simply because nobody tracks the right metrics.

There is no obligation for organizations to require absolute proof. A growing numbers of companies are reaping the benefits of contracting with a higher degree of trust.

As long as we continue not to measure the real cost of not trusting, to provide real evidence, the excuse to do nothing will persist with those organizations who do not see this as an opportunity to gain a competitive advantage.

The following observation is probably self-evident to most people, but it is included for emphasis.

There is significant inertia in the system created by the need for a safe approach rather than a smart one...

7.3 Contracts in a Context of Trust

I would argue that it is a self-evident truth that we work more effectively with people we trust. We communicate more openly and completely. We share problems and possible solutions. We are more likely to reduce the "checks and balances" that add overhead and other costs to the process. The converse is true too.

Contracts do not work well if we cannot trust the other party.

In the end, we trust people. This may be present internally within our organization or in our relationship with a buyer, contractor or supplier. The effect of trust or mistrust can be felt in any relationship or organization. If our shareholders or equivalent do not trust us, we need an auditable process. The same is true if we do not trust our shareholders. The auditability of work is costly, as it adds steps to work flow and likely adds time or approval requirements.

We can build a relationship with individuals. We can share experiences and learn about them as people. Once we have a personal connection, it is so much easier to stay in touch with them, relate to their needs and respond to their concerns. This works both ways. Although we may not return a call from a stranger, we will probably return a call from someone we like and whom we trust.

We trust PEOPLE rather than BUSINESSES or ORGANI-ZATIONS, because we connect with people.

People who share part of themselves (be it through a golf game, a meal or a common experience in raising children) are easier to get to know and trust. We are more likely to rely on such people than on ones we have not met or connected with. This is likely to be true of a tradesman, a doctor, an architect or a clerk. We build trust in different ways. The three trust types in the Color Trust Model can help explain this a bit better – or at least a bit differently.

We can rely on our relationships with people, and the better we know them as people, the more likely we are to feel that we can rely on them.

7.3.1 Competence Trust

An important part of selecting a contractor or supplier is to validate that the selected vendor has the capability to perform the work required and deliver what the buyer needs. This is so obvious that I probably do not need to even mention it. I do mention it, however, because it is not always done or necessarily done well. An engineering contractor, for example, may identify specific experts and specialists in a proposal. When the work is awarded, those specific people may not be available because other work or opportunities came up or problems needed to be addressed on another project.

Generally, when we are at the selling stage of a project, we have a human habit of promising to the limit of our capability. We want our potential client to see our best side and to gain a level of comfort (trust) in our ability to address his or her needs. There are as many ways of trying to do this as there are people in the business. Recently, I heard of a proposal that was sent out from one engineering company that included a very impressive resume for one of the specialists required for the project. Unfortunately, the specialist in question had died the previous year.

Conventional thinking in the procurement cycle requires us to select contractors based on their capability to deliver the goods and services that we require.

Competence, once validated, is just one aspect of the screening or selection process. Many of the people I have spoken to have indicated that they prefer to work with people who are easy to work with but may be slightly less competent or efficient than another who is difficult or awkward. This seems to be true of any team member, whether the team member is an individual or an organization, such as a supplier. A possible exception may be a situation in which the interaction with the vendor is brief and precise and when the product or technology is clearly superior to anything else. The brevity of the encounter and the relative advantage offered by the product or service in which one combination may be selected over another depend on many variables, including the past experience of the buyer.

 Do we prefer a vendor with a reputation for taking care but who is less competent than a vendor whom we know can do the work but who may not be trusted to look after our interests as well as the first vendor would?

We really need to know from where we are procuring competence. In construction projects, we normally expect the design consultants to provide design expertise. And the contractor provides the expertise in construction. What is missing is the COMBINED expertise that delivers the most appropriate construction methods, materials and process that produce the best result for the buyer. Other factors get in the way of achieving this. A personal observation is that this situation is based on human issues such as job protection, concern that we may be exposed to unnecessary risk, challenges to the established power hierarchy, changes in the "rules of the game" and so on. These are all issues covered by red trust.

The buyer's best opportunities to harness competence of its supply chain are compromised by human responses to change.

Addressing the technical issues of a project and the expectations of the buyer of the product or service being offered is, relatively, the easiest part of trust building. If we have trust in the vendor, we may have a lower level of quality assurance, or we may be more flexible in the interpretation of technical specifications. If we are uncomfortable with the supplier, we may be more diligent in our approach to quality assurance.

Another part of the issue surrounding trust in our suppliers lies in our perception of their ability to really understand our business and the way we operate. To overcome this apparent challenge, we tend to give our suppliers a solution to solve. We prefer to use our own judgment and expertise rather than rely on those of an outsider. The fact that the "outsider" may have more experience and may have seen better solutions in our competitor's shop seems to be something we do not like to deal with.

As suppliers, we may be in a position to offer alternatives and to negotiate the specifications. The opportunity to offer a better product or a more competitive solution than the one originally requested by a client is something that may be permitted in the process being used by the buyer.

Given that we need something to be done, we typically rely on specifications to set out the specific form of the solution we have defined for our problem.

Who cares what they do as long as
we can bill the client for their time?

There is often more than just technical performance at stake in selection of a contractor or supplier. Issues such as time to market, the need to develop new technology, performance and reliability of the end product are all factors that should affect the choice of a vendor. The challenge is often one of defining the process for selection, as we are now measuring more than just price. If we specify a completion date, then we expect the vendor to meet that date. Repeated experience with failure to meet dates and the associated disputes, claims and counterclaims suggests that there probably needs to be a way of assessing the probability of a particular vendor delivering on a promised date. In the end, a slightly higher initial price may result in a better final cost or a higher project value if the completion date requirements can be met or improved upon. In other words, we need to take a broader view than just whether a particular vendor can deliver what it had promised.

Just getting the job done is usually not enough. Other challenges, such as time to market and performance of the end product, demand that we add other conditions to our contract that address things beyond just competence.

7.3.2 Integrity Trust

Most commercial contracts are designed around establishing a set of rules for the parties to work with. These rules have a tendency to favor the party that wrote the contract. No surprise there. Furthermore, the rules are often ones that have evolved. And the evolution of these rules is based on past bad experiences. The end result is a contract that is based on mistrust that, in turn, is based on previous experience that has left someone feeling cheated.

Generally, we exhibit evidence that we have low integrity trust in our business dealings – possibly for good reason.

So, we have contracts that are based on a foundation of mistrust. If there is low integrity trust, then the safety net for the buyer is a "watertight" contract. Such a contract has a significant power imbalance built into it. This imbalance is normally in favor of the author of the contract. The power imbalance is deliberate and is designed to give the more powerful party the clout to negotiate after award of the contract. Typically, buyers feel vulnerable after having signed a contract. This is part of the antidote to such feelings of vulnerability.

Common law, by its very nature – having evolved based on past cases and decisions – offers a degree of equity to both buyers and sellers. This said, we need to remember that we should not confuse the law with justice or fairness. Between our contract and the legal framework of common law and legislation, we have a set of rules that provide a first line of defense against unscrupulous practitioners, be they vendors or buyers.

In the absence of integrity trust, we rely on the legal framework of common law and legislation as a first line of protection against unscrupulous practitioners.

There is a second line of defense in circumstances in which a regulated profession is involved. If we retain an architect or engineer, a doctor, a lawyer or any other professional, then the codes of practice of the profession offer additional protection to the public. The degree of protection varies from one jurisdiction to another. Variables include whether the profession is self-regulated or not and whether there is formal legislation, professional association bylaws, or simply a code of ethics that is not mandated for the profession, but is merely recommended.

The second line of protection we rely on is the regulation of professions and other specific service providers through accepted and, in some cases, legislated codes of conduct.

We have identified two potential safety nets that are built into our social and working system. For many, these are not enough when it comes to larger acquisitions. So we add specific clauses to our contracts to address whatever we are concerned about. Why do we need a belt, braces and a third set of defenses? Is this because we are paranoid? Or do we feel that we need to protect ourselves against the potential unscrupulousness of others. Integrity trust seems to be a rare commodity in many business dealings.

In the absence of any other protection, and sometimes in spite of such protection, we add contract clauses to provide a third level of defense against being damaged by a contractual relationship.

7.3.3 Intuitive Trust

If integrity trust is rare, and if we feel a need to define what we want done in explicit and precise detail, then what is left for us to worry about? We can provide detailed and accurate specifications. We can use a "watertight" contract. Yet, if a deal feels wrong, will we go ahead anyhow? I suspect that the answer depends on many factors, not least of which is how exposed we feel. This exposure may be very personal as a result of making the wrong decision or making a decision that is right but harder to defend than the obvious and traditional decision (i.e., selecting the cheapest supplier).

Should we do business with someone if the deal does not feel right?

Will you sign the contract BEFORE or AFTER Saturday?

As we learn more about intuitive trust, we are discovering that reliance on this type of trust can be a sign of professional maturity. Unfortunately, inappropriate reliance on intuitive trust is also a sign of professional immaturity. What is the difference? Appropriate reliance is the result of having a solid and relevant knowledge of what we are making a decision about; this knowledge is tempered by experience. Given a solid foundation of knowledge and experience in applying such knowledge, we process new information quickly. Some of the new information may not even be consciously absorbed. As a result, in the Western world, with its high proportion of lawyers and auditors and with a need for measured accountability and a tendency to try to assign blame, we have built an aversion to what we refer to as a "gut feeling," "intuition," or "a sixth sense". These are hard-to-explain responses. So decisions based on these responses are also hard to explain. This equates to being hard to justify in most business situations in which we are accountable for spending our shareholders' money.

  Intuitive trust is complicated. Part of the reason for this is that we do not really understand what happens when things do not feel right. Basically, we do not trust our "gut feeling".

There is no place for intuitive trust in conventional contracting. This is because there is no place for this type of trust in conventional management practices with most organizations.

In informal discussions with numerous practitioners, I have noticed that many of the best managers, project managers and business people make decisions based on intuition. However, they tend to take care to validate their responses – sometimes after the commitment has been made. Also mentioned in these discussions is that, on those occasions in which the appropriate business processes or procedures led to a counterintuitive decision, the decision turned out to be a wrong one. In the absence of real empirical data to support any conclusion, the following statement is more of a gut feeling: Intuition is underrated in today's business world. Where speed to market, efficiency, cost-effectiveness and other values are of such importance, it seems strange that our traditions and need for full auditable accountability have become barriers to performance. Rather than just looking for better results, our practices lead us to caution and even to making bad decisions that we can justify rather than good decisions that we cannot.

A word of caution: Intuitive decisions (not raw emotional ones) can frequently be justified based on assessment or analysis, if or when we have the time. What we need is the right mechanism for explaining our decisions to others.

 We need to supplement intuitive trust with auditable practices until we have developed whatever other business skills are needed to allow us to be more effective in this trust arena.

7.3.4 Balancing Trust Types

Section 7.3 provides a quick look at trust types in the context of how we develop contractual relationships today. The first observation is that we have found substitutes for trust in the way in which we assemble contracts and manage the process of selecting our contractors and managing the contract afterwards. A second observation is that we still need to feel that we can rely on achieving our objectives (competence trust) and doing so in a business-like and fair way (integrity trust). We also know that most people prefer to work in a setting where they feel right about their relationship (intuitive trust). We need a bit of each of the three types of trust. More than this, we need to have a degree of balance in these types of trust. If we cannot achieve the balance in other ways, we know we can achieve some measure of trust in the contract document to protect us where we feel exposed.

Like most things in life, trust needs to be in balance. The right balance for a particular situation depends on many variables that define that situation.

We studied contractor-selection processes and how they may be improved. We interviewed many practitioners in the business of buying goods and services. One of the most interesting findings from this study was that most of the interviewees took into consideration their intuitive trust in the process of selecting a contractor or supplier. Perhaps more interesting was the observation that they were open to the idea of developing better ways of selecting vendors to reduce the incidence of claims, underperformance and other contributors to cost and schedule overruns.

Are we ready to use our intuitive skills in building a contract strategy? Indications are that the best contract administrators already do.

7.3.5 Adapting Contracts and Managing Them Based on Trust

As noted, there is a cost associated with lack of trust and with inappropriate allocation of risk. There is, therefore, a real opportunity to gain a competitive advantage in any business in which there is a supply chain if we can improve our purchasing skills. Trust appears to be playing a critical role in the next evolution in business practices.

Trust-based contracts offer huge potential for better working relationships and, therefore, for better competitive advantage.

The first reaction of most people to "trust me" is NOT to trust whoever used that phrase. Generations of declining trust in business dealings are hard to overcome. And they will not be overcome overnight. Are trust-based contracts even likely? A recent survey conducted as part of the overall research to better understand trust in business suggests that it is already a reality in practice. This study is part of Ramy Zaghloul's doctoral research, so is still unpublished. It does indicate that a significant percentage (likely as much as 50% or more) of the work done in capital construction projects is done outside a formally documented contract. The elements that make up this type of work include the following.

- Contracts are formalized and signed after the work has started or even after it has been completed.
- A "new" contract is started after an old one is finished, but neither a change order nor a new contract is formally created.
- Change orders are instigated when the work is done before the required changes to the contract are made.

Initial inquiries suggest that there may even be a higher incidence of such informal arrangements in maintenance arrangements, emergency work and the use of professional services (e.g., lawyers, doctors, other professionals). These types of arrangements rely on the creation of a contract in common law and on the good will of the parties engaged in the work as much as, or even more than, any other documentation does.

These types of informal and essentially trust-based contracts survive, it would appear, because both parties are somewhat vulnerable, and their reputation and opportunity for future work or access to expertise or services depend to a large extent on the residual relationship after the contract.

 Trust-based contracts do not happen overnight. Most of us will need time to adapt from generations of no trust-based contracting.

Why are people getting into these acts of unprotected commerce? Maybe the additional time and effort and the extra cost of protection in the form of a contract are just not worth it. Possibly the barrier of the contract and its imposed behaviors make the relationship less sensitive to circumstances. Specifically, contracts appear to offer an exchange of more protection for less responsiveness and speed. Today, so many of our projects are susceptible to changes, be they imposed (e.g., environmental issues, political changes) or voluntary (e.g., cost-saving ideas, technology changes, schedule acceleration). Whatever the drivers, there seems to be a significant amount of work being done by contractors for buyers without a formal agreement in place. And the users of such contracts are reaping rewards in terms of flexibility, responsiveness, speed and cost savings. The real question seems to be: Why are we not using trust-based contracts more widely? Or more visibly?

Early adopters of trust-based contracting are starting to reap some benefits already.

Possibly, part of the challenge in adopting trust-based contracts lies in the fact that they are not recognized as being what they are. Possibly, the label "trust-based contract" has an inherent warning built into it.

Maybe we need a more acceptable label. "High performance" or "super contract" may do the job because neither mentions trust – to which we seem to have a negative response because of our built-in aversion to what it may imply. I like the term "SMART Contract" for cost-efficient, responsive and fast contract formation.

And what are these implications? Perhaps the most important of these is the exposure we have in trusting someone else, as we give up part of the control of the business we are supposed to be managing or protecting. Any demonstration of trust, such as the wording of our contract, the process we use for selecting a contractor or even the administration of the contract, will normally make us vulnerable to the other party. And is our aversion to such vulnerability not the reason we use conventional contracts in the first place? And are we willing to continue to pay the premium associated with such traditions?

The first part of trust-based contracting requires that we demonstrate a certain level of trust through our contract documents, the processes we use for selection of a contractor and administration of the contract itself.

It would appear that the informal use of trust-based contracts is not based on a prescriptive approach. The use of such contracts is usually driven by some overwhelming need. The thing that the need overwhelms is the requirement for a formal contract to exist, initially at least. Thus it makes sense that the use of trust-based contracting is based on need. If this is true, then the absence of structure in the form that individual contracts of this nature take is inevitable. This tells us that there is unlikely to be a specific, or even a recommendable, approach to formation of such agreements between parties. Chapter 12 explores the use of this type of contract. SMART contracts require a bit more effort and intelligence on the part of the people involved than do the traditional methods.

There is not likely to be a magic formula for successful implementation of better contracting practices, as the approach will need to vary to suit the unique and special needs of each organization.

We have looked at the role of trust and seen that it adds a layer of complexity. Potentially the payoff can be significant. One of the simplest things we can do about trust is to build and maintain it through our contractual relationships and one way of doing this is by administering contracts consistently and fairly.

Chapter 8

THOU SHALT NOT MESS UNDULY WITH THE CONTRACT AFTER IT IS AGREED

> *Administration of contracts:*
> *payments, meetings, and documentation.*

The contract administration process is largely dependent on the relationships developed between people. Strict adherence to contract terms is often not the case, as people work together to implement what they believe is the intent of the contract. Actions, however, may change the effective terms of a contract, so not following the terms may invalidate them, with resulting complications if things turn sour later.

Chapter 8 looks at key issues that need to be considered in establishing the way in which contracts are administered and the implications of specific types of action by a party to the contract. Chapter 8 does NOT offer any strict interpretation of clauses in the eyes of the law. Nor does it offer likely outcomes of any litigation. The situations that govern the outcomes of legal process vary too much by jurisdiction and circumstance. Finally, Chapter 8 does not attempt to tell anyone how to run their business. Everyone has their own particular needs when it comes to administration of a contract. How the paper flows, what approvals are needed for payments and changes, and other administrative details are left to each organization.

Chapter 8 is more about the potential and most common pitfalls in managing contracts than about how to manage routine administration of contracts.

8.1 Payment and its Implications

One would think that getting paid is a good thing. This is generally true provided that we are paid for the right thing and provided that neither making a payment nor receiving a payment violates or changes the terms or conditions of the contract we are working under. Most of the time we get paid for goods and services rendered in such a way that the contract is kept whole. However, should the contractor's subcontractor or supplier ask us to pay them directly for some work, we may end up having a new and possibly separate contract with that subcontractor. We may also have a dispute with the contractor, who may have lost some commercial leverage as a result of our payment and may not be in a position to encourage or coerce the supplier to do specific work under the terms of its contract with us. We will have interfered with the management of the contract and may well inherit all of the problems associated with such action.

Payment is a major factor that determines intent of a contract and just who has a contract with whom.

It generally pays to read invoices. If we pay an invoice for items that are not included in the contract, then we may have added those items to our contract without a change order. This is fine, provided that we want the additional work and the paper trail is acceptable to our organization. Equally, invoicing and getting paid for something that is not in our contract may add more than just cash to our coffers. For example, if we were a construction contractor and we billed – and got paid for – some engineering work associated with a redesign of part of what we were building, have we inherited professional liability for the design? And how much of the design will our bit of work have affected?

Even getting paid may have some implications that can fundamentally change our contract.

8.1.1 Payment Processes

The flow of money is to a contract as the flow of blood is to the body. Stop it, and the animal dies. Just about any business has constant outflow of money: payroll, rent, financing charges, payments to suppliers and subcontractors, and so on. Inflow is dependent on the ability to collect money from our clients. This requires a set of good billing practices tied closely to the good will of the customer. To this end, a deliverables-based approach is useful. Break down the work into manageable pieces, with associated processes, that we can deliver within a billing period. Deliver the work, get the buyer to accept it, and get paid for what we have done.

One of the engineering companies that we tried this approach with went from complex billing processes, with submission of all sorts of backup material, including time sheets and invoices for photocopying, to a few sheets of paper listing the drawings and specifications produced and delivered to the owner. This list included the all-inclusive price for those items based on a set of "rules". The rules themselves served to eliminate a lot of rework and scope creep. This was done by defining what a complete drawing was. In the case of a drawing, the process was set out as part of the definition of the deliverable. The process included only one review by the client. Because only one review was permitted, the owner took extra care to identify all the concerns they had – and a specified turnaround time too – and the number of design iterations (and associated irritations) was reduced. As the pricing was based on historical records, the consulting engineer was able to turn a greater profit (fewer iterations) and the client saw faster progress and a more organized delivery of the engineering work for the project (fewer irritations). Everyone came out a winner. In this particular case, receivables were reduced to a 35-day delay versus the historical 55- to 70-day delay that the company was used to. This meant that the amount of cash flow that was financed was reduced, and once again, profitability increased.

 Most contracts clearly set out the process for getting paid, and they identify what the contractor is entitled to. This is particularly important to the contractor, as it affects cash flow.

There are some things that we should not expect payment for. And there are other things for which payment is deferred. We should not expect to be paid for our own errors or omissions and the work required to repair them, as they are our responsibility. If the error is initiated by a fault of the other party, then we are entitled to payment for making good the situation. If we volunteer to do additional work or provide additional goods, then we should not expect payment if there is no agreement – real or tacit – to be paid for so doing.

If we are entitled to payment, then we need to be aware of the legal or contractual requirement of the buyer to hold back part of the payment. In North America, for example, the lien act in its many guises often requires the buyer to hold back a percentage (often, but not only, 10%) of payments owed a supplier of goods or services to an improvement. This money is deemed held in trust against the failure of the contractor to pay its suppliers. The same principle normally applies to a contractor in its payments to its suppliers. If the legislation of the legal jurisdiction we are working under does not cover such a requirement, the contract itself may. So read the contract. If there is a conflict between the terms of the contract and any applicable legislation, the legislation will normally apply. So know the relevant law too.

It is important to understand what we will NOT get paid for. It is also important to understand how long it may take to receive payment.

8.1.2 Holdbacks and Lien Acts

Holdbacks and the lien act are not simple issues, and it pays to understand the underlying principles. Because the specifics of each lien act differ by jurisdiction, the following observations are somewhat generic. I make no apology for this. If an apology is due, it is from the legislators, who believe they have the right answer – why else would we have so many different "solutions"?

The first point is that, in the United States and in Canada, one is normally required to hold back payment on any improvement to a real property. Let me translate. Improvement means any work done or goods provided to a project that result in enhanced value. Real property normally means a tangible fixed asset, usually real estate is involved. This is normally a product of construction, but it may also include things like a car (sometimes covered by a separate mechanics lien act). It would be interesting to see if implementation of a large piece of technology such as a corporate software system, would be considered an improvement under the meaning of every such act in the land! In most cases, the answer is clear if the technology is in the form of a building or pressure vessels or even in the form of a modification to a set of programmable local controllers. However, as we move into software and knowledge-based improvements with little or no physical presence, such as an improved robot control system tied to a Manufacturing Resource Planning system – is this covered by the lien act? And if so, is this consistent across jurisdictions? Within the scope of this book, I merely want to bring attention to this situation, and I hope that future contracts and products do not get caught on the wrong side of an interpretation of the law.

 In the United States and Canada, the buyer is normally legally obligated to hold back part of a payment to a contractor for an improvement on a property.

If the lien act does apply to our contract, then we are normally considered to have held back the required amount of money from our contractors, whether we really have done so or not. What does this mean? There are two important parts to this. First, in most of the lien acts in North America, there is a provision that says one cannot contract out of the act. In other words, if we decide to bypass the act through words in the contract or through our actions, the rules will still be applied as if the contract had included the act and its provisions. A common provision is that the lien act will deem us to have held back payment from a contractor. In the event that we have paid the contractor and some event triggers litigation against us, the courts will assume that the money is still held by us, and we may end up paying twice for the goods and services covered by the holdback. Just how much this amounts to will vary from one jurisdiction to another. Typically, however, the amount of the holdback is 10%, although some jurisdictions look for up to 15%.

 The amount of money to be held back is dependent on local (usually state or provincial) legislation.

Federal government projects typically follow different rules.

Each jurisdiction may have other legislation that could affect our contract and how the contract may be dealt with in the event that it runs into trouble. I say "runs into trouble" because in most cases our contracts are not interfered with by the courts until we invite the courts into them by resorting to litigation to solve our problems.

There may be other legislation in the area in which we are contracting that we need to be aware of because it may create rights and obligations that need to be incorporated in the way we administer a contract.

There is a saying that possession is nine tenths of the law. So maybe it is a good thing to hold back money from a supplier. If nothing else, people argue, this is a way of providing supplementary financing for our business. The unanswered (and often unasked) question lays out the issue of what this is costing us. The cost is hidden in the price and it covers the cost of financing the work that the vendor has not been paid for.

Perhaps a better reason to hang on to money owed to a vendor is to apply pressure to get what we need, but have not yet got, from them. A debt catches the attention of any business worth its salt. This is a good way to do business, provided always and only that this is a justifiable and ethical process. The offset in this approach is that the vendor loses trust in us as a buyer. If we do not rely on the vendor in the long term, then this may not be important. If, however, we are likely to need that vendor – and possibly the vendor's network of suppliers and partners – then the relationship needs to be considered in our business decision. If the contractor is trying to perform, but is failing for good reason, holding back payment may actually further damage their capability to meet our needs. The moral of all of this: We must understand what we are doing when we hold back payments to our suppliers. We probably need the trust of our suppliers in the long run.

If we hold money that may be due to, or claimed by, another party, we have some anxiety and obligation associated with that held money. If we have no right to hold it, we may have a liability as well.

The contract will typically set out the process for paying or getting paid for goods and services provided under it. It is not uncommon for payments to be made periodically, such as monthly. These payments are intended to help the contractor maintain cash flow while protecting the buyer from having to pay for work not yet done. The latter is particularly important to buyers who are borrowing some or all of the money to cover the cost of the project. From a contractor's perspective, it is reasonable to expect payment in accordance with the terms set out in the contract. Yet often, owners will build in a process that extends payment beyond this defined period or provides extended times for checking and processing the required documentation. Timely payment of progress invoices from a contractor is important for a number of reasons. Profitability alone is not a bad reason. The following calculation is an oversimplification of reality, but it serves to illustrate the point:

Contract value = $10 million; duration = 10 months. Contractor can borrow money at 1% per month.

Contract terms require the buyer to pay monthly progress payments (of $1 million per month) within 30 days of the end of the month.

The contractor has priced its work as follows:

Cost of the work excluding overhead	$ 9,000,000
Interest (1% for 1 month x 10 months on $900,000 cost per month)	$ 90,000
Overhead costs ($60,000 per month)	$ 600,000
Contingency	$ 110,000
Profit	$ 200,000
TOTAL	**$10,000,000**

If the buyer makes payments in 90 days instead of 30 days, as required by the contract, then the net additional interest that the contractor needs to pay will be on the $900,000 per month that is delayed by 60 days. This works out to $180,000 if we do not include compound interest. This amount is very close to the total profit. Furthermore, the contractor's line of credit may be stretched beyond the limit set by the bank and will therefore trigger additional problems. The problems, in an extreme, will trigger the process that leads to bankruptcy. Typically a contractor caught in this situation will defer payment to its subcontractors and suppliers wherever it can. Guess what? Now all we have done is push the problem down one level in the supply chain, where it may well trigger a similar reaction . . .

All that this does, in most cases, is create additional operational problems for the people we are trying to get better performance out of.

Finally, don't be quick to criticize the small margins that I have suggested in the example. General contractors in North America earn about 2% on turnover as a fairly typical profit before tax on an annual basis. At least, this is what it was the last time I was able to get hold of this type of information from sureties, banks and other sources that I consider fairly reliable. Contractors with this type of margin have little room for error. The reason they can make a good business out of this is that the investment is relatively small compared with turnover, so the return on investment is leveraged. These ratios do, however, help highlight the importance of cash flow to the average contractor.

Progress payments are due to a contractor as set out in the contract. Holding these payments as leverage for performance or for some other reason can, and should be, dangerous.

8.1.3 Lender and Investor Concerns

Enough projects are financed in whole or in part that it pays for us to understand some of the more important concerns of lenders and how these concerns are likely to affect the contract itself and its administration. Lenders have an obligation to the people who own the money they are lending. This is the depositor in the case of a bank, or the current or future pensioner in the case of a pension plan. A lender will, therefore, try to be as confident as possible that any money lent is returned, with interest. This may well mean that the lender to a project will require specific assurances. They may need assurance that the loan amount will never exceed the value of the underlying asset. They may need some assurance that the project can produce revenue at the rate that will meet loan repayments. Whatever the specifics, the general issue is that a lender wants reassurance that the loan will be secure. The more secure the loan, the more competitive the interest rate and any associated fees and charges will likely be.

From time to time, we are obliged to add terms and conditions to our contract that affect how money is released to our contractors, because we have lenders or investors that first need to release the money to us.

In addition, the buyer who has obtained the loan will likely look for similar assurances that help protect its interests. These needs, from a business perspective, are often reflected in contract clauses. The difference between a buyer and a lender in this situation is that the lender is less concerned about the final cost of the project than the buyer is. Lenders need to protect an investment that is based on a guaranteed rate of return (as specified by the interest rate on the loan). The buyer has to turn a profit on the asset by making it produce net revenues that repay the loan and provide a real return on their portion of the investment. Their risk is greater. It is not unreasonable for them to expect a greater return on their investment. They will be looking for assurances that their investment will provide a return.

One of the biggest drivers for specific conditions being imposed by lenders and investors is the need to protect the investment.

The sidebar "A Healthy Investment" illustrates how the specific concerns of the lender affected the contract and the contract price. (See another story with a different emphasis in Section 3.3.3.) Other common requirements of lenders and investors include the following:

- Payment to the contractor should always be less than or equal to the current value of the asset. This is especially true if non recourse project financing is used. In this type of financing, the only recourse that the lender has is to seize and liquidate the asset that is being financed. They do not have recourse to the borrower's other assets. One outcome of this is that payments may be slow; this would be reflected in the additional financing costs that the contractor has to carry.

- A third-party review of progress payments is required. This often goes hand-in-hand with the previous requirement. The objective is that there is an independent review of progress payments to protect the interests of the lender or financier.
- Performance of the buyer (as opposed to the contractor) may be required. Part or all of the associated risk may be passed onto the contractor through specific clauses in the agreement.

A Healthy Investment

A health care company wanted to build a new senior citizens' home. It owned the land but had no capital available to fund the construction of this new facility. It was not able to negotiate a loan from a bank.

A design-build corporation offered to design and build the new seniors home but required a commercial deal that would allow it to obtain the required project financing, then sell the whole facility to a life insurance company that invested in sound long term real estate deals.

To allow the life insurance company to invest, it required a predetermined and assured return on its investment and assurance that it would get paid. The assurance was provided by the regional health authority who underwrote the project and guaranteed its operation and lease payments for 20 years.

The lease was established based on a 20 year amortization of the building cost. The land was sold for $1 to the pension fund, which underwrote the capital loan from a bank to pay for design and construction. The bank released funds to the designer/builder and was repaid by the pension fund on completion as the pension fund could, under its mandate not finance construction projects. At the end of the 20-year lease, the health care company would buy back the land and building for $1.

Clauses added into a contract may add significant cost to the project. This will have a big effect on the profitability of the final outcome.

Often, the money provided as a loan to a buyer for the purpose of paying a contractor is based on the value of what has been built so far. This may require a third-party assessment of the work-in-place to be part of the payment process.

8.1.4 Implicit Contract in Common Law

Section 8.1.1 looked at the effects of paying someone else's contractor, which can lead to a significant change in the terms of one or more contracts. It is important, generally, that we follow the rules we set out in the contract. If we do not, then we may change those rules or terms through our actions. In other words, we need to match the terms of our contract and the way we administer it.

Just making a payment to someone can create a contract that may be implied in common law.

Another important principle to follow is to avoid managing the work of our contractor. There is a fine line between doing this and providing direction or determining compliance with the terms of the contract. For example, specifying a way of doing something or a sequence of work is on that line. Generally speaking, if, as a buyer, we require a specific approach to the work, we will be responsible for the efficacy of that approach or sequence. If we did not include this specification on the bid documents, then such a specification will normally trigger a request for a change order to cover any additional cost and time associated with this requirement. If the requirement is the result if an intervention by the buyer's staff overriding the management of the contractor, then the buyer normally assumes responsibility for the consequences of such actions.

We should be able to see that there can be a very fine line between providing direction and managing the contractor's business. Stepping over this line as a result of how we manage the contract invariably leads to additional risk and responsibility falling to the buyer.

We must be very careful when we feel the need to tell the contractor or supplier how to do something.

Actions can speak louder than words. If we decide to do something contrary to the terms of the contract, and both parties appear to comply with this change, it may become permanent.

8.2 Notices Under the Contract

For the past decade, my research team and I have been looking at project management and what works and what fails. In that time, I have specifically requested examples of failed projects for us to look at. All have failed for one ultimate cause: a breakdown in communication. Communication is key to the success of every enterprise, and management of high-performance contracts is no exception. Many contracts recognize this by specifying how to formally communicate.

In other work, I have found that the most common cause of project failure is a breakdown in communication. Thus, anything that helps us communicate more effectively has to be good.

8.2.1 Purpose of Notices

Most of us do not like surprises. Consider a very simple contract to get your car serviced. You take it to the service station early in the morning and they tell you it will be ready at noon and the service will cost you $100. At noon you come by, and they tell you it is not ready yet. They also know that it will cost more, but they do not tell you because they think you are upset enough already. Later in the day, on your second visit, you collect your car and are presented with a bill for $150. You are not too happy.

Now consider a slightly different situation: the service station notifies you of the problems during the morning and obtains your approval for the extra work and the higher bill, saying they thought it may take until the next day and will cost $200. In this case, you would probably have been delighted to get your car the same day and only had to pay $150!

 No bad surprises please. People do not like them!

The purpose of the requirement for formal notices is based on the need to maintain communications and to be able to provide an audit trail for these communications. If something needs to be communicated that may affect the management of the contract, its terms or compliance by one party, there is a need for all parties to the contract to communicate this. Informal communication that may or may not have been effectively transferred or an absence of formal evidence of the communication having taken place is what we are trying to avoid. Why? When it comes to acrimony, our memories are selective, and what the other side may remember may not be what we recall as fact. This requirement for formal notice helps establish the rules for evidence in the event that we need to have recourse to legal action.

 The more time we have to prepare, the more opportunity we have to respond most effectively. It is always good for us to know what is going on.

 Formal notices provide an auditable set of evidence of what who said to whom and when – just in case we need it!

8.2.2 Notices Often Defined by the Contract

Read the contract and specifications. The rules – if they exist - will be set out in the documents that have been signed. In the absence of formal rules set out in a contract, it pays to make sure that communications that are likely to be needed because they affect the contract in some way are delivered in writing. Ideally, they are delivered in a way that provides evidence of the delivery. If necessary – and it is good practice to do this anyhow – make sure that there are designated people with whom to communicate.

Know the Contract. If we do, we will be able to manage it more effectively.

It is not uncommon for notice provisions to include standard documents and other required communications set out in the contract. Notice through regular communication can, for example, include changes to a schedule being notified through issue of a revised schedule. If the recipient (who typically will have required the updated schedules on a regular basis) does not pick up on the changes, then they may well be liable for the consequences of having effectively ignored a formal notification. Other ways of communicating that may be acceptable as formal notice under the contract can include minutes of meetings, issues of modified drawings and specifications, requests for pricing of changes or the responses to these requests, and most commonly, correspondence.

Out here is where we get the final client requirements.

Notices can be delivered under the terms of the Contract in ways that may at first not seem obvious. Read all material, including schedules, shop drawings and specifications, with care.

There are two views of the indirect approach to delivering notices under the contract. The first is that we need to be aware of these so they do not cause us problems later. The other is that they can be used in a way that may effectively disguise the formal notice in a larger document or in one that is a distraction from the real message. The caution implied in what I have just said is that less scrupulous people will deliberately use these approaches to help prepare for future claims. Even the most scrupulous of practitioners will look for implied or "hidden" notice in such documents in the event that they are cornered by the opposition in a confrontational situation.

Sometimes people like to disguise a message because it is bad news for someone and messengers with bad news have a tendency to get shot!

In my doctoral studies, I was able to observe that people tend to demonstrate some reluctance to let others know of a problem because they are concerned about souring the relationship before it is absolutely necessary to do so. In my research I sought a response to the statement: "Contractors save claims until the project is complete or almost complete because they do not want to spoil their relationship with the (a) Owner and (b) Consultant". The responses indicated that 43% agreed or strongly agreed in the case of owners and 54% agreed or strongly agreed in the case of consultants. The study was based on opinions of expert practitioners in the Canadian construction industry. In addition, 37% disagreed or strongly disagreed in the case of owners and 34% disagreed or strongly disagreed in the case of consultants. Although the vote was split, there is still a strong feeling that there are hidden claims, in this industry at least, that remain unannounced because of concern over relationships with another party to the Contract. In response to another question, 97% of respondents agreed or strongly agreed that many contract disputes are known about (by at least one party) for a long time before they are dealt with.

 Sometimes, people like to be as nice as they can for as long as possible. So we keep bad news hidden until we have to expose it.

8.3 The Power of Progress Meeting Minutes

Well-run projects have regular progress meetings. Managed well, they can be very effective in management of a contract. People do play games even here, so we need to understand how we manage meetings well and how meetings can be used to improve contract administration.

8.3.1 Using Progress Meeting Minutes to Serve Notice

Unless the contract is explicit about limiting the ways to provide formal notice, minutes of a meeting can serve this purpose. Minutes should be an accurate reflection of what was discussed and agreed to at a meeting. If the parties to the contract are present at the meeting, and if the minutes are accepted by both parties, then the minutes represent witnessed documentation of what transpired. Arguably, this is one of the best-documented forms of evidence of agreement (or disagreement) between the parties.

 In most cases, it can be argued that anything in a set of meeting minutes, from a meeting at which both parties to a contract were represented, constitutes a formal notice under the contract.

Because meeting minutes provide documentation of a discussion, it is important that the minutes be accurate. If the minutes refer to the Contract and use specific terms defined therein, then it is important that there is agreement on the meaning of these defined terms. For example, the Contract may define the term "Subcontractor" to mean one specific subcontractor that the contractor is expected to use for a specific and specialized part of the work to be done under the contract. The meeting minutes may refer to the Subcontractor in relation to a specific incident or a request for a change in the work. If the intent of the writer is to refer to the specific subcontractor defined in the contract, then the minutes are correct. If the intent is to refer to a different subcontractor of the prime contractor, then the minutes are incorrect. Conventionally, the defined term is capitalized, although when used generally, the term is not capitalized.

Once again, the message is the same: read and understand the Contract before using it to manage the project.

 Pay special attention to the specific wording of the Contract and the meeting minutes.

Meeting minutes become particularly important if they record the result of a discussion or a decision that affects the terms and conditions of the Contract. If the meaning of the minutes that refer to a change are ambiguous, and the result of the ambiguity is a dispute, then we have acquired an additional problem we did not need. Life is difficult enough without avoidable misunderstandings. Generally, based on my experience and that of others I have talked to on this issue, the longer the ambiguity remains undetected, the worse the result of the discovery will be if there are, in fact, two or more interpretations. It seems that our expectations will harden over time if they are not managed.

Just as clarity in contracts is important, so is clarity in the wording of meeting minutes.

8.3.2 Using "Standard" Minutes

Confirming that nothing is wrong at each meeting is a useful way of drawing out truth and creating time to address situations in a more organized and structured way. An interesting way of doing this is to pre-write part of the minutes of regular progress meetings. This sounds strange at first blush, but it does make sense in a perverse sort of way. The sorts of things to include in the pre-written part of the minutes are the issues that experience tells us are perennial problems. Here are some examples:

- The Contractor confirmed that it has not been delayed by late issue of drawings or any other information required from the designer.

- The Owner confirmed that there are no concerns outstanding that may delay or reduce payments due to the Contractor.
- All parties feel that communication is open and effective.
- The work to be done over the next week or month is understood and all issues of coordination between parties have been addressed.
- There are no concerns over safety issues for the work to be done before the next meeting.

There is some equity in these statements. They are not aimed at just one party, but identify the concerns that each party may have. How can we develop this standard part of the minutes? I recommend that this be done at the first meeting. Each party should contribute the reassurances they need to hear if they are to feel comfortable that the more common problems are being monitored in an organized way that will minimize the risk of a future problem. How does this reduce subsequent problems? Simply, it raises the main issues of concern on a regular basis. If there is a problem, then it is incumbent on the person who has a concern to raise the issue before the record shows that there is no problem. Use of a scale that runs from 0 for bad news to 10 for perfection. This allows some people to score each item developed by the project team and start some discussion over issues that may lead to subsequent problems.

Some project managers and contract administrators use a set of pre-written minutes as a guideline and agenda for progress meetings.

Progress meetings provide a real opportunity to improve the outcomes of the project and to reduce risk and conflict in management of the contract. How? By adding elements to the pre-written part of the minutes, we can highlight issues that need to be addressed and managed on an ongoing basis. These items should not be focused only on technical issues, but should also address the business aspects of the contract (e.g., schedule, payment, extras, commercial problems, supply chain issues). In addition, there should be some discussion about what I call social issues. These are those factors that affect relationships. They include an assessment of relationships with third parties and the public, openness of communication, morale of the team, attitude and more. This list should be based on the experience of the team.

It pays to use the regular progress meeting to test the waters on technical, business and social issues.

Dr. George Jergeas, who has done a lot of fascinating work in the area of dispute resolution and partnering, has developed an approach to management of alliances and partnerships in a project that uses many of the concepts we have identified (Jergeas and Cooke, 1997). Jergeas got his inspiration from the chapter on partnering in "Project Management: The Managerial Process" by Gray and Larson (2000). In addition, he suggests that a facilitator should work with the parties to a contract to help them manage relationships. He uses a number of tools, one of which is a set of questions – developed by the team – that address the issues that can lead to problems on projects. He does this by asking each of the parties three questions, then getting them to find ways to address the issues that are identified by the process of answering the questions. Here are the questions he asks:

- What is it that others do that caused us problems in the past?
- What is it that we do that could have caused problems to others in the past?
- What can we do to improve on these experiences while we work on these projects?

(These questions are not direct quotations.)

Part of the regular progress meeting process can include a questionnaire that addresses relationships and other key "soft" issues.

The pen is mightier than the sword. Managing the minutes by writing them is an approach that provides a degree of control over what is documented as history. If we write the minutes, we have the first say. More important is the fact that we will know what is in the minutes even before they are issued. If we do not write the minutes, then they are new to us when we read them. We will need to take extra care in reading them to be sure that the facts recorded therein are accurate and that any opinions or concerns we had are not just recorded but are precisely represented. Too many people do not bother to read minutes of meetings, and, over time, anything that was recorded incorrectly becomes fact.

 One of the most powerful ways to manage the outcome of meetings is to be sure that we have control of writing the minutes.

Responding promptly to items we disagree with in the minutes of a meeting that we attended or at which we were represented is absolutely critical if we want to preserve our position on an issue. If we fail to respond and the document audit trail shows that we had a copy of the minutes or that we should have known what transpired, we will likely be deemed to have agreed with the accuracy of the record (in this case, the minutes of the meeting). If we disagree with the minutes of a meeting – or any other document – we must make sure that our disagreement is documented well.

If we disagree with what was said in a set of minutes of a meeting, it is vital that we let the writer of the minutes know. And we must do this in writing.

A good progress meeting formula includes two main ingredients. The first is a consistent and forward-looking agenda. The second is a short part of the meeting dedicated to administration of outstanding issues. How do these ingredients fit together to make success more likely?

Progress Meeting Agenda:

Topic	Time Required
1. Completed work since last meeting	1 – 5 minutes
2. What was missed	1 – 5 minutes
3. Resulting problems and how to resolve them	1 – 20 minutes
4. Administrative issues to date	
[Here we discuss the issues that are important to the parties and which form the "pre-written" portion of the meeting's minutes.]	5 – 15 minutes
5. What needs to be done next week in order to be on time?	1 minute or less
6. Problems associated with achieving next week's targets	2 – 10 minutes
7. Resolution of problems and issues identified in item 7.	2 – 30 minutes
8. Other Business	0 – 15 minutes
9. Recognition and learning from past week's activities.	2 – 10 minutes.

Minimum time for meeting*	15 minutes
Maximum time for meeting	2 hours

* The minimum times are achieved when all is on target and there are no major issues. We achieve this by planning well, communicating effectively, resolving issues off-line wherever possible and helping each other achieve the target objectives from last week.

The second item we need to pay attention to is embedded in item 4 of the proposed meeting agenda in "Progress Meeting Agenda". Here, there are issues that are important to the effectiveness of the team. They were identified early in the process when we identified the ingredients for the "standard" minutes. The business of meetings on projects is discussed more in "Don't Park Your Brain Outside".

Consistency and efficiency are key to any regular meeting. The best meetings are short, are focused and emphasize coming work rather than assignment of blame.

8.4   Documents that Affect Contract Terms and Conditions

When we are angry or emotional, many of us like to vent. It is best not to do this in writing, but if we do, we should not even consider sending anything before the third draft. Why? Anything we put in writing is something that can be discovered in the event of litigation. Our credibility is at risk in a court if we can be made to appear to be emotional, unprofessional, or in some other way exposed. Conversely, contemporary documentation can help provide useful and objective support through documentation of decisions, questions and answers, clarifications, actions and other things that reflect the management of the contract by all of the parties involved.

Filed paper is important, as it is real and often contemporary evidence. This can be good and bad. We need to make sure our paper is good.

Now, let us take a look at the more common forms of paper accumulated in the course of managing a contract.

8.4.1 Memos to File

I am going to start with an example of what I consider to be the worst type of paper documentation on a contract, although I admit to having resorted to it myself in the past. I have always considered memos to file as being a weak way of addressing problems. That said, it is a legitimate form of documentation that is often used today. A memo to file may serve a useful purpose if it is professional (as opposed to vindictive), factual (as opposed to opinion) and appropriate that it be sent to a file. Ideally there is a third party or at least a second person involved, as they can then corroborate the document was contemporary as opposed to being added at a later date.

If there are concerns over raising an issue that may create more problems – perhaps because of timing or because it may be the last straw that breaks the camel's back – then send a memo to file. If one party is negotiating the resolution to one issue and does not want to muddy the waters, this may also be an appropriate way of addressing the short-term need for documentation on another issue.

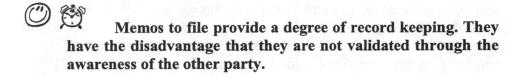

Memos to file provide a degree of record keeping. They have the disadvantage that they are not validated through the awareness of the other party.

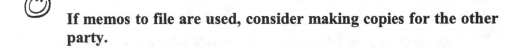

If memos to file are used, consider making copies for the other party.

Memos to file should be factual and ideally have an audit trail for subsequent validation or verification.

8.4.2 Diaries

Over the years, I have found that keeping a diary helps. There are some basic principles to follow here, though, to ensure that the document produced has maximum weight in the event that a dispute has to go to court. The first of these principles is consistency. A professional or project diary, if used, should be consistently maintained. A diary that has the same format and is kept current on a day-to-day basis is more robust than one that only records events that suit a particular perspective or objective. The consistency of a diary should extend to several aspects: content, style, format and, if appropriate, the level of detail recorded. If the diary is kept current each day or on some other regular basis, then it will be relatively easy to demonstrate that the record is contemporary. A contemporary record is usually considered more reliable than one made some time after the event, as time tends to distort facts.

The contemporary nature of diaries serves a useful purpose, provided they are regularly maintained and consistently completed.

Part of the professionalism of a diary lies in consistent use of an objective and honest style of reporting that either keeps to the facts or separates fact from observation, speculation, or opinion. One other point on diaries before we leave the subject. I have found that it also pays to have a diary format that is tamper-proof. The format I use is a hardback book with numbered and bound pages. I start each day with a double-ruled line and then state the day and date. I leave one blank line between points. I use the left margin to highlight points or to "tag" them with an icon or other symbol. I use the right margin for afterthoughts. Not even I can tamper with my own diary. This adds credibility to the document as a source in the event of a dispute.

A diary should be kept in a professional way. Remember that it can be used in court in a worst-case situation.

8.4.3 Written Communication

Nowadays we have many media for written communication. Some forms are safer than others. Typically, the more tactile, the more likely it will be recognized as legitimate later, as it will be harder to falsify in any way. The importance of this becomes really clear when we realize that the value of written communication is that it is harder to refute later.

If something is in writing, it is clearly harder to refute later.

If we receive communication that we disagree with, we should make a comment to that effect as soon as possible. Remember, it is easy, later on, for the other side to refute that we made a verbal comment objecting to the written one that they had sent. In this situation, we should not send a note to file objecting. It is important that all recipients of the original communication, especially the author, receive a copy of any written disagreement, correction or clarification. This clearly establishes our position at the right time in the history of a contract and preserves our integrity, position and possibly our rights under that contract.

If we get a written communication that we disagree with, we should confirm our disagreement in writing and send it to the author of the first document.

As discussed in Chapter 1, the author of a contract is normally held liable for the consequences of ambiguity in the resulting document. So, too, in most cases we will be held liable if we produce a written communication that can reasonably be interpreted in more than one way. It is important to read what we have written – and to do so objectively! Consider getting someone else to review a letter or other written document if it may be sensitive or critical to the success of the contract.

Watch for ambiguity. The wrong interpretation may be used against the author of an ambiguous document.

A common feature of contract clauses that deal with formal notice provisions is that the notice is delivered through the mail or through some other third party. The reason for this is that this action provides an audit trail for sending and receiving such documents. This clause is one that I now read with interest if the contract document was prepared by the other party. This is because my sense is that this clause is a pretty good indicator of the trust level that the other party operates under. The more formal and bullet-proof the clause and the notice delivery method is, the lower the general level of trust that the other party works under is likely to be.

Any critical document needs to be conveyed to the other party through an independent third party.

8.4.4 Confirmation of Important Items

The more important something appears to be, the greater the attention we probably should pay to what is going on. This is simply good business practice in today's world.

Even if something is in writing, it sometimes pays to confirm what we have understood.

8.4.5 Other Forms of Communication

In today's wonderful world of technological wizardry, we need to consider the place of other forms of communication. Table 8-1 looks at types of written communication and I offer my humble opinion as to what to do about this type should an important message be sent via that medium. If we agree with the message, we may wish to confirm our agreement. If we disagree with the message, then we need to communicate that as well.

Communication Medium	Response to Important Message	Rationale for Format of Response
Faxes	Confirm by fax and mail.	Faxes can be generated on a computer and can incorporate electronically captured images. These can be manipulated. So a hard (paper) copy of faxes should be kept. In a low-trust situation, consider sending back a copy of the fax with an acknowledgment letter or note to the originator.
E-mail	Acknowledge e-mail and confirm by letter or keep a hard copy.	E-mail can be manipulated electronically. Some images change depending on printer settings and other hardware configurations and limitations. Be aware of these. If in doubt, print what is critical and file it. In low-trust situations, return copies of the printout and interpretation to the originator.
Web page	Print any screens. Note dates and if there is any contractual implication. Confirm our understanding by letter.	As Web pages change from time to time, the version we looked at needs to be recorded. Shared project information, coordination and project office information may be updated moments after we download information we consider current. This may lead to wasted effort or other losses. If necessary, we should agree on timetables or electronic notices of changes to contract-critical Web sites.
Other electronic communication	As for Web pages	The same rationale used for Web pages and e-mail applies here.

Communication Medium	Response to Important Message	Rationale for Format of Response
Handwritten note or message	Copy and handwrite confirmation of our under-standing on the note. Return a copy with the note added.	Typically, we place lower emphasis on handwritten instructions that we do on more formal-looking documents that spring forth from printers, for example. Yet many detailed instructions are given by handwritten messages.
Graphics/ Drawings	Acknowledge receipt and provide comment on any real or implied contractual implication to what has been changed.	Drawings can be complicated. Sometimes changes made on a drawing are not clearly identified in the description of revisions. In low-trust situations, we should confirm our understanding of the changes made. Alternatively, we can agree that only changes identified in revision notes will be considered legitimate or relevant.
Verbal	Confirm in writing. A multipart carbonless form works well, as we can keep a copy for the file and one for the person who has to implement the change.	Verbal instructions are often convenient. They are easily misunderstood. They are harder to remember than written ones. It pays to have a log of these instructions and clarifications, as they can survive memory fade, staff turnover and other things that make them transient. The date and time of such instructions are automatically recorded if we confirm verbal instructions correctly.

Table 8-1. Response media to important communications.

The extent to which we pay attention to how we document communications and provide an audit trail largely depends on the level of trust that exists and is justifiable in any given contractual relationship. In the end, we need the audit trail if we ever have to resort to litigation. Consider adding to the agreement the protocols for communication.

Written communication can be much more than just material delivered in writing. We need to be sensitive to the various forms of documented and timed information that we send and receive and to the value of such material in our management of contracts.

8.5 Legal Recourse: Will the Real Contract Please Stand Up?!

What have we really been talking about in Chapter 8? Basically, if there is just one important message, it is that the contract really is the <u>beginning</u> of subsequent negotiation. The real contract is what we subsequently apply. The original contract represents the ground rules. The changes to the contract need to be understood. If we have a dispute and need to understand the contract, we will typically find it is based on the original signed document (if there is one)... and a lot of "Aahhh yes! But . . ."s!

The original contract is just the starting point from which we will have negotiated the real thing.

Our behaviors and the changes we make after signing the contract can – and do – lead to added complications by potentially modifying the terms and conditions of the contract. What does this really means to us in a practical sense?

8.5.1 Reading the Contract

This sounds too silly to even mention, but I will: Read the contract. I have already said this a few times throughout this book. So what is the big deal? I have a few real examples of projects in which it seems people did not read the contract before signing it.

Read the Contract!

CASE 1:
A CEO was finalizing a contract in a foreign land, and was trying to catch his imminent flight back to his home country. There were a few minor changes to be incorporated. What the client offered was to make the changes while he checked in and went through immigration and security and then would meet him at the departure lounge so he could sign and take away a copy of the contract. The CEO of the supplier organization jumped at the generous offer to help him catch his plane. The client turned up with two copies of the contract just as they were boarding the plane. The supplier signed both copies (already pre-signed by the client) and took his copy with him. The changes made by the client were not the ones agreed by the two in their discussions. The supplier now had a serious problem!

CASE 2:
A contractor called on a claims consultant and requested that the consultant help him prepare a claim under his Stipulated Price Contract with one of his clients. The consultant read the contract and pointed out that it was a Unit Rate Contract, and the contractor did not have a claim. The contractor had not read the contract – even by the end of the project!

CASE 3:
When a technology transfer contract failed because the buyer was unable to meet its obligations to pay the technology supplier, it became obvious that the buyer had not read the contract it had signed. It thought the payments were discretionary!

We could go on forever with such examples . . .

 The first step in good contract administration is to read and understand the Contract!

If we are new to contract administration, or we are working in a new field, a new legal jurisdiction or with a company we have not dealt with before, we should be extra careful in trying to interpret the Contract. Meanings of words, how local custom works, how the courts in this jurisdiction interpret the law and other fine points can help muddy the waters. If there is a critical new element in the contract arrangements, such as – but not limited to – the ones just listed, we should consider getting a third-party opinion from someone who has the knowledge and experience that we are short on. That said, I have probably worded this paragraph badly. I have likely not helped a professional contracts administrator with lots of experience by implying that they do not know their job. One can choose to ignore this advice, although I still follow this advice after 30 years in the business and a decade of research. It simply does not pay to ass-u-me!

   We must make sure we really do understand the Contract – even if we need to get a third-party opinion to verify our understanding. It will save a bunch of time and trouble later.

8.5.2 Knowing What the Contract Means

It usually pays to get good professional advice. We have made the world of contracting and buying more complex than it probably needs to be for most situations, and we have built in a large transaction cost in the process. This said, we have what we have, and if we are certain what the rules are, it is a good idea to talk to and get advice from the people who know what is going on.

Good professional contract administrators, procurement specialists, buyers and lawyers will have seen and lived through many contracts in their careers. They will normally be able to assess the strengths and weaknesses of a given position.

If the original Contract is the starting point for subsequent negotiation, then there is more to the Contract than this original document. Change orders, memoranda of understanding, re-issued drawings and specifications, correspondence and other records may also affect the interpretation of what the intent of the Contract is. This certainly makes life more complicated. Consider the following examples. In all cases, assume that the Contract is silent on the points that are raised unless specifically stated in the scenario.

- The Buyer pays after 90 days, but the Contract requires payment within 30 days. The Vendor does not complain or ask for interest until this has gone on for more than 6 months. Is the Contract deemed to have been amended by the delay in payment by the Buyer and the tacit acceptance of this change by the Vendor?
- A Designer, working for the Buyer, makes a change and issues a revised drawing or specification. The Contractor makes the change. Later, the Contractor requests payment for the change. Is there entitlement?
- A Vendor proposes a different product to that specified in the Contract. There is no response from the Buyer. The Contractor implements the change. The Buyer accepts the end product. Then the Buyer seeks a credit from the Vendor for the change. Is there any entitlement?
- In minutes of a meeting, the Buyer asks that the project be accelerated because the current forecast completion date is no longer acceptable. The Vendor requests additional payment. The buyer's representative puts a note on the file, without copying the Vendor, saying that the Vendor should do this for no additional cost. The Vendor finishes a week early and submits a request for payment for all premiums (e.g., overtime, extra shipping costs, additional planning, supervision). The Buyer refers to the note on file and says no payment is going to be made. Is this the right decision?

Clearly, we cannot tell from the limited information what is right and what is not. There will be many other factors to consider. My point is that these other factors exist. The Contract and its interpretation will depend in part on people's behaviors, how all parties appear to have treated the terms of the Contract to that point, precedent on the project, and more.

The interpretation of a Contract is not always straightforward.

Looks like the notes of the post-tender
negotiations are being delivered.

8.5.3 Last Resort: Turning to the Law

First, do not get confused. Using legal advice is appropriate long before turning to legal recourse. Any lawyer who understands contract law and practices in the relevant jurisdiction will provide a useful and, hopefully, objective view of the case and how solid it may be. The trick is to fully brief the legal advisor. We tend to tell them a story that is heavily biased toward our view of the case. Try to get someone to play devil's advocate for the other side. Maybe our case is not as strong as we first believed it to be.

Turning to lawyers is not the same as turning to the law.

A good and honest advisor with a complete picture of the case and a sound understanding of the opponent's case will be able to advise well. My experience is that this advice will be not to pursue legal action unless we have a very strong and clear-cut argument or we really want our day in court, regardless of whether we win or lose. Either way, if we do decide to go ahead, we must remember it will cost us in the end, and it will take much longer than we think.

Before turning to the law, we should be sure that we have a good case.

If we do resort to legal action, we must remember that the cost can be high.

I cannot stress it enough: be sure that the lawyer has a complete and balanced picture. I have been down the path to litigation on behalf of the company I worked for or for a client too often not to see the pattern. We start with what appears to be a good case. But as we go through the inevitable phases of negotiation and examination for discovery we discover the weaknesses, flaws and other problems with our case and the strength of the opposition's case. We can only hope that the opposition is going through a similar discovery. Then, maybe we can settle before we get to the courthouse.

Make sure that the lawyer has ALL of the relevant information, not just what we want him or her to hear.

Every cloud has a silver lining. If we are able to resolve the dispute, we may gain a friend and ally out of the process. Chapter 10 looks at dispute resolution in more detail.

A well-resolved dispute can be a real opportunity to build relationships with the other party. Try looking at a dispute as just such a way of doing future business.

In Chapter 8 we looked at some of the main issues we encounter in administering contracts. One issue in particular warrants further attention: changes. Let us now look at managing these contractual changes.

Chapter 9

THOU SHALT DEAL RATIONALLY AND FAIRLY WITH INEVITABLE CHANGES TO A CONTRACT

> *Changes and Markups: Types of changes, processes, costs and games played.*

Changes to contracts are inevitable. The potential for disagreement in the pricing and impact assessment of contract changes is very high. There is irrational behavior in how we deal with contract changes. Some owners and consultants take requests for additional payment from a contractor personally. Others are affronted by such requests. Some contractors play games with changes.

A study that tried to shed light on the behavior of contracting parties investigated the markups associated with changes (Semple, et al, 1994). Contracts in many parts of the world specify what the markups for overhead and profit should be. Often such markups are specified – interestingly, they are frequently set in steps of 5%, which is somehow redolent of arbitrariness rather than science.

The study uncovered some interesting things. First, nobody seems to agree on what the markups for changes should be or even what is included in them. We found industry-wide disagreement on what a direct cost is and what should be included in indirect costs versus overhead or profit.

Chapter 9 presents the findings of this study. It then examines the way in which contractors, subcontractors, owners and others respond to changes, pricing strategies and just a few of the games that people play with changes to survive in business, cover errors and omissions, make a better profit or otherwise achieve some specific objectives.

There is a bit of mystery attached to markups and what they cover.

9.1 Change Orders

Change Orders in many contracts develop because we have a genuine change or because we forgot something when we prepared the original contract documents. The former is the result of someone coming up with a better, or at least different, idea to solve a problem. The latter is often the result of haste in preparing the documents. If this is the case, then the trade off in management terms is between the advantage of getting started and the need for absolute accuracy at the outset. We discussed this concept earlier in Chapter 3 when we looked at contract strategies. Another reason for changes is simply having forgotten things. The forgotten things are often, and depressingly, ones that we forgot on previous occasions. Or they are things that we did not spell out in the contract documents well enough. There are many examples of these kinds of change orders that lead to problems. The problem is built into the process we adopt in the first place.

For example, if you are a contractor trying to win a stipulated price contract based on having the lowest price, are you likely to include for items that you know should be in the contract but are not explicitly stated as being required? Typically not, otherwise you might price yourself out of the deal.

The problem arises when we request the inevitable change order. The person responsible for preparing the bid documents is now exposed. They may appear to have left something out and this may be construed as negligence or an error or omission. The normal human condition is to protect one's self; if this were me, I may argue to the contractor that he or she should have included this item as it was obviously required to achieve the desired result, citing the normal requirement that obvious discrepancies need be brought to the attention of the owner prior to bid. Now I have a dispute rather than a change order, if I am not careful.

Why did I choose to explain this at the beginning of Chapter 9? Because there seems to be emotion attached to change orders in many situations. My little story serves to illustrate one possible reason for the resistance of some contract administrators to want to address change orders as being normal occurrences that require normal processes to manage them well. Part of the process needs to include management of client expectations. Going back to the role of the consultant, I should take the time to explain the risks associated with a reduced time or an inadequate budget for completing the design. The client can then allow an appropriate design contingency to match the situation. If the client knows that this type of change order will be required because the design consultant was asked to reduce the time or effort to produce drawings, there is no need to get concerned or emotional when it happens.

Changes on contracts come about for different reasons. They are a normal occurrence, and they need to be managed appropriately to avoid unnecessary conflict.

Really, it is complete.
All that's missing is in the change orders you didn't approve.

9.1.1 Change Basics

What is a Change Order? Most contracts permit the buyer to make changes to the scope or quality of the end product that they have commissioned from a vendor, even after the order has been placed. This is to allow for the inevitable improvements and modifications to a project that will be needed over time. Errors or omissions may drive the changes, but changes are more likely to be initiated in response to other factors. These other factors can include market conditions, innovations in technology, modifications to suit unknowns that could only be guessed when the contract was awarded, necessary changes to address actual circumstances or working conditions, changes driven by regulatory or other compliance requirements and more.

The process of issuing a change order generates a contract within the original contract. The items that can be changed, the rules for pricing them and the general conditions that apply to management of the changes are often set out in the terms and conditions of the original contract.

A Change Order is like a contract within a contract. The terms set out in the original contract will typically apply to the Change Order.

Allowable changes can span a broad range of items. The rules that normally apply are based on common law. One of the important principles in common law is that any change in the contract that goes to the root of what the contract is about may nullify the contract. As a result, this principle has often been turned into a more codified and manageable set of rules in commercial contracts. One such rule sets limits to the scope and nature of changes. For example, it is not uncommon to see quantities change in a contract. Changes beyond 15% of the originally specified quantities may result in renegotiation of unit rates or other terms of the contract that are then considered to be rendered invalid by such an adjustment. The percentage variation permitted in a contract may be specified and can vary considerably.

What is allowed to be included in a contract through a Change Order may be dictated by the contract itself.

9.1.2 Processes Needed for Managing Changes

In any contract that may require changes, it is prudent to include a vehicle to accommodate them. Incorporating changes into a contract simply modifies the contract to include the changes. This is based on the presumption that the parties to the contract agree on the basis under which the changes are made. If the parties cannot agree, then the contract may have a provision for forcing the contractor to perform the changes with resolution of the inevitable dispute after the work has been done. Let us look at this in a slightly different way.

First, if there is agreement, we effectively have all of the ingredients required to create a contract in itself under common law. We have an offer, acceptance, an intent to be legally bound, an exchange of value and, presumably, legality. In other words, we have a contract within a contract. If we fail to agree, then we need to have a mechanism that allows work to continue and allows the differences between the parties to be resolved in an equitable way, usually after the work has been completed. To make this happen, we really need to agree on the business mechanics for dealing with these situations. The best – and most obvious – place to do this is in the general conditions of the contract.

Changes to contracts are common, so some agreed approach to management of the changes that are likely to happen should be assembled between the buyer and vendor at the outset.

Particularly when the buyer has a need to be fully auditable, the mechanism for managing contractual changes will be set out in the buyer's standard contract terms and conditions. The processes and procedures set out in such terms and conditions are normally difficult to circumvent. This is good and bad. The good part is that the auditability and the required checks and balances are preserved. The bad part is that, especially today when time is so critical on most of our contracts, we delay rapid processing of changes and the necessary paperwork as we grind through corporate procedures that are too unresponsive for the situations that they now need to deal with. Many of the processes that we deal with in this arena are old and based on old ways of doing business. Chapter 12 looks at this in more detail, examining how to work in a faster, more flexible way that better reflects some of today's business conditions and realities.

Formal arrangements to address all changes are common, especially when there is a strong need for auditability.

Formal mechanisms provide a high degree of comfort to participants, as they cannot be rushed. Furthermore, they usually make assignment of responsibility for a decision harder to trace. This is the result of multiple approvals and intermediate steps that dilute decision-making and accountability. Many business systems acquire this attribute over time, as different people get blamed for a decision and then introduce new steps to reduce the risk of repeating this unpleasant experience in the future. This is likely the source of much bureaucracy, the subject of another debate.

I think he'll make a fabulous change order administrator.

To counter the less flexible and, therefore overly slow processes, we see more informal arrangements being created. Often these informal ways of getting work done in spite of the contract and the processes we are supposed to follow are based on people knowing and trusting each other based on having worked together before. These informal processes are invariably simple, based on an agreement to follow the formal process later and a reliance on trust in achieving an overall equitable resolution to address cost, schedule and other changes under the formal terms of the contract.

The risk in this kind of approach lies in staff changing before matters are finalized, interruption of the process as a result of some external influence or a political shift that changes the underlying rules. There are other risks too, but they revolve around being able to sustain the informal relationship through difficult circumstances and still protect the interests of both parties to the contract. I only recommend this approach when the intent of the contract and the interests of the contracting parties are preserved at all times.

There is an advantage to having informal change order mechanisms in place, provided that the interests of the parties are always protected.

9.2 Process for Managing Changes

There are many different ways of managing changes in contracts. No one way is necessarily better or worse than another. The real test is that the parties to the contract are happy with the process and that the process does not interfere with the efficiency and effectiveness of the contract it is trying to support. Let us look at a generic flowchart for managing changes to a contract.

9.2.1 Flowchart

The flowchart in Figure 9-1 shows the main steps that would generally be required to create a change to an existing contract.

The flowchart in Figure 9-1 is intended only as a generic approach.

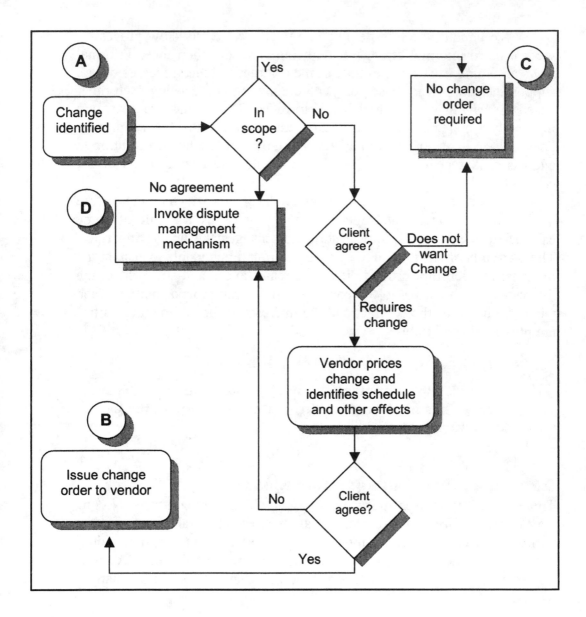

Figure 9-1. Change Order flow chart.

In the simplified flowchart in Figure 9-1, we start at point A, where a change is identified. There are three possible outcomes. At point B, we have issued a Change Order. At point C, we have decided that we do not need a change order after all and at point D, we have a dispute of some sort. Typically the dispute will be over whether a change is really in the scope of the original contract or whether it will affect the cost or schedule of the project.

 Simple or complex, the change order process should address all of the real issues involved in effective management of the contract.

I have not had the chance to formally research this point, but it would not surprise me that the ease with which a change order is processed has much to do with how money is managed in the buyer and seller organizations. If approvals are required at the board level, then people probably feel that they are more exposed and require more care in processing and validating every aspect of the request for additional money. If the decision-maker is directly involved in the change and has the authority and the funds to proceed, then the decision is likely to be made more quickly. This is probably true of both buyer and seller organizations. In a buyer organization, attention is paid to the increase in cost of a particular project. In a seller organization, the focus is on profitability of the project and sustainable relationships. Clearly other factors also affect how changes are managed, including the profile of the project, who the sponsor is, what the priorities are, the economy and more.

The further away that money decisions are made from the day-to-day management of the project, the more complex the change management process is likely to be.

If we can see a connection between where a change is approved in an organization and the relative added complexity as the decision moves away from where the action is, then it is clear we need to understand the way that organization works. This is not a new problem, and most practitioners have been there and have found innovative solutions for managing the imposed limitations that they have to work under. This was discussed earlier, in Section 3.1.1.

The story of payment in "boulders" in Chapter 3 is one of many I have heard over time. All of the stories have a few things in common:

- The people involved know their part of the business well
- They want to do a good job
- They try to understand the business objectives of the enterprise they work for
- They are creative
- They are confident in their own capabilities and their ability to demonstrate that they are working in the best interests of their employer.

In the end, the concerns of senior management and the direction of the enterprise we are trying to serve come first. What is important to the enterprise is what should dictate our approach to managing the business. Change Orders are just another part of the business. The way we manage funds and sanction payments on contracts should be aligned with what is important to the organization. When this is not the case, we can only hope that the people in charge will take corrective action to achieve the right result in spite of imposed bureaucracy.

In most cases, what is important for effective management of changes is dictated by how funds are managed and sanctioned.

9.2.2 The Need for Process to Meet Buyer and Vendor Needs

What Chapter 9 advocates must look like anarchy to some. This is certainly not my intent. I advocate intelligent management. This means that we need to be sure that the real interests of all parties are protected in the processes we set up. If bureaucracy hamstrings us, I encourage us to look for work-around solutions. That said, we should always make sure that we and our business partners (in this case, the other party to the contract) protect our own and each other's interests. If this holds true, then we need to have some process for managing changes to a contract that works for both the buyer and the seller. This process should ensure that the technical and commercial needs of both parties are protected. To do this, the change management sequence probably needs to be tailored to the specific circumstances of the project. For example, if the project is time-driven, a slow and precise procedure for managing changes, which will potentially delay the completion of the work, is inappropriate. That same process, in cases in which the concerns are over cost and meeting technical performance, may be absolutely correct.

 Whatever the process for managing change, be sure that the commercial and other needs of all parties are considered.

9.3 When is a Change a Change?

A contractual change that is not seen as a change by one party must be a dispute. In fact, the converse is just as true. This means that recognition of what is and what is not a change is of particular importance if we want to avoid disputes.

If we do not recognize a change for what it is, it will end up as a dispute if pursued by the party that believes it is entitled to compensation.

9.3.1 Recognizing a Real Change

A material change to a contract affects the cost or time required to perform the work under that contract. There is also another type of change: one that in some way alters the performance of the end product of the contract. Performance has two dimensions: quality and scope.

If we want to be able to identify a change to the contract, we need to define the basis of the contract thoroughly, clearly and concisely. The most common tool for doing this is the Work Breakdown Structure (WBS). The WBS was developed originally as a graphical and hierarchical way of depicting the scope of work for a project. Detail increases at each lower level. Well-designed WBSs are entirely based on deliverables, starting with the project at the top. A modification to this tool is the SMART (strategically managed, aligned, regenerative, transitional) Breakdown Structure™ (SBS). The SBS differs from the WBS in that the first two layers are not deliverable items that are required for the project. The first level in an SBS is the mission of the project. The second layer describes the Key Results that tell us the project mission has been achieved. Associated with the Key Results are the stakeholders who are concerned with them. The third layer describes each Key Result in terms of what is to be delivered. Each subsequent layer describes these main deliverable in more detail.

This approach to breaking a project (or a contract) into defined pieces helps to describe the quality and scope of what we are trying to achieve. If one of these elements changes later, it represents a change to the project or contract and can thus be readily identified. In the absence of such a structure and standard approach, we are left guessing and arguing in too many cases over whether we have a change or not.

The WBS or SBS can be prepared such that, when it is read in conjunction with the contract and any specifications or drawing that are embedded in the contract, it is straightforward and identifies the true scope of work.

Changes to a contract should be easily determined if the scope and quality expectations are clearly articulated in the first place. Unfortunately this is not always the case, so changes are not always clear. A Work Breakdown Structure or, better, a SMART Breakdown Structure™ goes a long way to help address this.

There are different types of changes. We should take a quick look at some of them. Deletions to the original scope of work represent one type of change. If we delete an item, it needs to be taken out of the scope of the work. Associated with this item will be a cost. The cost is not necessarily the same as the price in the contract, as the price may include for fixed overheads and other project-related costs that are spread over many items that are to be provided by the vendor. Sometimes a buyer will anticipate the possible elimination of these items from a contract and will ask for a price in advance for deletion of such items. This process represents the credit that the vendor will give the buyer if the buyer deletes the item. There may also be time restrictions associated with these deletions, as the credit for deleting something that has already been manufactured, shipped and installed at the buyer's site will clearly be much smaller than the credit for deleting it before it is ordered or manufactured. In fact, in the former case, there may even be a "negative credit," as the item may be specially manufactured and will therefore not have a ready market. De-installing it and returning it will also have an associated cost. The total may well exceed the cost of keeping it.

Any deletions to the work to be done or goods to be provided under a contract generally constitute a change, unless the terms of the contract provide for such a deletion.

Additions to the contract are the next obvious type of change. Additions may be easy to recognize, because they result in an additional component being produced or delivered. There are also situations in which the addition has more to do with process. For example, the buyer may place restrictions on the vendor to suit a new union agreement, to coordinate with another contractor, to react to complaints regarding noise or dust from neighbors or for a myriad of other reasons. These restrictions may upset the sequence of work or the efficient delivery of goods or services by the vendor. In such a situation, the vendor is normally entitled to recover the additional cost and may be entitled to an extension to any contractual completion date.

Some additions are harder to measure and defend. A common problem built into agreements for consulting services is the apparent open-endedness of the work. This is probably a good thing from the vendor's perspective if it is paid based on time and materials (cost plus). If, however, the contract is based on a stipulated price, this is quite dangerous. Many consultants make this mistake. Consider a fixed fee for design of a building. The buyer is presented with a design and requires changes. The building is redesigned. The buyer has new ideas and requires more changes. The building is redesigned again. When does this stop? How much redesign is part of the original contract? In this type of situation it makes a lot of sense to define the process: We will design once, the buyer will review the design and we will make necessary changes. After that the buyer can make as many changes as it likes, but all of the changes are extra to the existing contract.

If the buyer adds to the work to be done, either in terms of quantity or difficulty, then the contractor is normally entitled to recover the additional cost and time for such extra work, plus a reasonable allowance for overhead and profit.

If the changes result from modifications to the original scope or quality of the work or are the result of imposed sequencing, delays or other constraints not initially contemplated in the contract, then there needs to be some compensation to the contractor for any additional effort, time or other resources required.

On one project, a data field was added after the database tables had been defined, the coding completed and the data entry screens built and tested. It took 3 months to add this one data field. The cost exceeded $1.2 million. Had the data field been identified and included at the outset, the additional time and cost would not even have been measurable. On another project, the buyer asked what it would cost to add a cutoff valve to one process train. The buyer was told that the cost was insignificant, as the basic engineering and plant layout were not complete yet. The request for the change came about a week before commissioning was due to start. The valve was ordered and installed. The delay to construction and commissioning, however, was about 8 weeks. The cost was estimated at more than $1 million in direct costs and several tens of millions of dollars in operating losses and lost revenue. It pays to understand the effect of time and progress on a project has on a change. At some point, a change becomes prohibitively expensive.

The real effect of any change will typically increase over time.

9.3.2 What Can Change?

We have seen that changes affect schedule, cost and performance. (Remember, performance is defined as a combination of quality and scope.) And we have seen that changes to the contract normally allow for the affected party to be entitled to some compensation. Finally, we have noticed that if one party to a contract requires and initiates a change, then the other party is entitled to recovery of the cost and time associated with any effects of that change.

Section 9.3.2 looks at some cases that illustrate these principles in a bit more detail. Fortunately, there are endless examples of these types of changes, so finding examples is easy!

Following are five cases to illustrate changes under different headings. Arguably these headings are somewhat random, as we may debate their relevance and appropriateness. I really do not want to go there. The headings are merely there to demonstrate that some sort of classification of contract changes is possible.

 There are different types of changes. In the end, they all have some sort of effect on a contract.

Note the way we can improve management of these changes by defining some part of the work more clearly. The following examples are taken from engineering and other professional services agreements, as they can be used to illustrate most of the sins we commit in defining the nature of the work to be done.

Case 1: Scope of Work

In this case the client asked for the design and construction of a warehouse to accommodate pharmaceutical chemicals, both as a raw material and while a drug was in "quarantine" between manufacturing stages. Part way through the design the client added a request to accommodate a change in the manufacturing process. This change entailed replacing rack storage of one product that was delivered in large plastic bins with bulk storage in silos. This simple request required considerable additional engineering to accommodate the need to be able to trace the product throughout all manufacturing stages. The material in the bins would need to flow in a way that avoided the mixing of batches as they were delivered and used in production. To do this, mass-flow bins were required, with vibration capability to stop rat-holing and bridging in the fine granular raw material that was to be stored there. To add to this scope change, the material was to be moved in the warehouse and in the plant using a pneumatic handling system. The air that such a system uses has to be specially filtered and processed when entering the system and has to be treated to remove suspended fine particles before being exhausted. The buyer considered these changes, and modifications to the building to accommodate them, to be included in the basic engineering price. The price had been quoted as a fixed fee. The additional work had to be separated and explained before the client agreed to pay the difference. This was an *addition to the scope of work*.

Case 2: Change in Working Conditions

In another case, a reconstruction of a historical building was to have taken 12 months. After award of the contract, the neighbors complained about noise before 9:30 a.m. and after 4:00 p.m. They also did not want noise on the weekend. This essentially cut down the working time to 6 hours per day for 5 days per week. The union contractor had to pay for a minimum of 8 hours per worker per day. Not only was the working time per day severely reduced, but the net cost per hour was increased by this one change imposed on the contract after the terms and conditions had been set. This was a *change in working conditions*.

Case 3: Deletions and Additions

In another project, someone made a change in the ceiling tile to be installed. The request came after a high-rise building had ceiling tile delivered to each floor via an external hoist, which had then been removed to allow the curtain wall contractor to finish enclosing the building. To accommodate the change, the existing tile had to be taken out of the building and the new tile brought in. The contractor decided to use the two freight elevators for this task. This affected the warranties on the elevators, affecting another trade. This was a *deletion* of one tile *and addition* of another. Reshipping and restocking charges also applied.

A change in the scope of the work inevitably changes the cost and the time required. Such a change may also affect the methods to be used and may thus affect warranties and other contractual terms and conditions.

Case 4: Unit Rate Change

A change in the quantity of concrete required for construction of aprons on a new harbor seemed trivial. The change involved the reduction of a 10-inch unreinforced concrete slab to a 6-inch reinforced slab. This 40% reduction in concrete affected the contractor in several ways: First, the availability of ironworkers for cutting, bending and placing the rebar was low. This meant that the project would be delayed, resulting in additional cost to the contractor to maintain site services, trailers, supervision and other costs. Second, the concrete had been ordered through a ready-mix company, and the reduction brought the total to be purchased from this supplier to below the minimum at which the contractor got the price break that the bid was based on. Finally, there were overhead costs assigned to the concrete rate. The cost of labor increased significantly to add the cutting and placing of the reinforcing steel. There was no labor saved in forming and finishing the concrete, and this part of the labor cost had to be charged on significantly fewer cubic yards of concrete, thus increasing the rate again. These costs could not be recovered easily without a direct charge being made in the pricing of the change order. The owner expected a savings or – at worst – no increase in price. The contractor expected its significant additional cost to be recovered. This change resulted in a *change in unit rate* for concrete in slabs on this unit rate contract.

> **A change in the quantity of one or more items to be provided under a contract affects the distribution of overhead and profit, may affect sequencing of the work, and through this, may affect the unit price of each of the affected items.**

Case 5: Delays

For a change of pace, let us consider this case, where the buyer incurred an additional cost. On a new project, an owner had decided to purchase some of the long-lead items. One of these items was the structural steel for a series of transit warehouses. This delayed the foundations and other work by about four weeks. There was no available storage on site. So the steel had to be kept on trailers in the shipper's yard. Demurrage charges were considerable. In addition, there were other delays earlier, although not as long as the one that led to these additional costs. The site was not available to the contractor at the outset, resulting in a 2-week delay. This delay pushed some of the foundation work into a period of unusually bad weather, which the contractor argued led to further delay in completing this work.

> **Timing is important in the delivery of goods or services. A change in timing or sequencing affects availability of key resources, commitment to other customers, inventory usage, financing of large or expensive components and more.**

Just about every project has standards of quality set out in the various contracts that are awarded for the completion of the work. If more than one contractor is involved, then the buyer is normally responsible for the quality of the work on which a subsequent contractor operates. For example, a tile setter was given a schedule and a set of specifications for installing wall tile. This was expensive marble tile, and the finish was not simple. The tolerances were very fine for the tile setter. The concrete and drywall that he was given to work with did not meet his exacting standards, and he was required to do a lot of additional work to prepare the surface for his tiles. This led to a request for a change order.

When setting standards or expectations for one part of a project, we need to be aware of the sequence of work to be done. We need to add supervision and quality assurance at all of the preceding steps to be sure that there is no need for rework or for additional work on the part of later trades or vendors as a result of what we have accepted from a previous trade or vendor.

Quality of the end product is affected by many factors. Quality assurance costs grow exponentially as quality standards are raised.

What we consider to be quality is best described as getting what we expect or better. Arguably, the key to quality is to ensure that the parties to a contract are aligned in their expectations of the goods or services to be provided. Another view is that the definition of performance is the extent to which the end product achieves the intent, or a change in the intent, of the buyer.

One well-known corporation in North America ordered electronic components from Japan in the 1980s. The buyer expected quality to meet at least North American standards and was reluctant until that point in time to trust Japanese products – likely a hangover from the '50's and '60's. The buyer was very careful in his specifications. He would only accept shipments in which five in each 1,000 components failed to work. To be sure that it was understood, the buyer pointed out the clause in the contract and then reiterated this point when the deal was finally signed. When the first shipment was delivered, there was a small package included with the first 1,000 parts. It contained five more parts and a note. The note said, "Please find enclosed the five nonfunctional parts required by your specifications."

I may not have all the words in the note correct, but the message was clear: Why ship <u>any</u> parts that do not meet the real specifications? All too often we set suppliers the task of solving our solution. We consider it none of their business to understand what we are trying to achieve with our project and how their contract fits into the overall objective.

> **Performance and expectations go hand in hand. Often, contracts are defined in terms of the solution to a problem rather than a request for a solution to a problem. The end result is that the measure of success of the contract may be based on the wrong set of assumptions.**

9.4 The Effect of Changes and how it Alters Over Time

All changes affect the way we work. For a start, we only like changes that we feel we have some control or influence over. A change to a contract does not take place, as we have seen, until there is agreement to the existence and validity of the change. It is not uncommon for there to be several changes to any one contract. If this is the case, there may be a cumulative effect on the performance of the contractor as a result of these changes. On occasion, there are opportunities created by one change that can actually help with another change to the same contract. There are several ways that changes can affect the vendor. We should look at a few of the more common ones here.

> **Changes need to be identifiable. Then they need to be recognized and accepted by all parties. Finally, we need to be aware that there are synergies between changes that may have greater – or fewer – effects than may first appear.**

9.4.1 Effects

There are both direct and indirect effects of changes on the vendor. The direct effect is the direct cost and time that are saved or added as a result of taking out or adding work or scope to the project. The indirect effect is harder to define and measure. It includes reassignment of overhead, disruption to the planned flow of work, inefficiencies, overtime and lost productivity, effects on team morale, and more.

The effect of a change can be assessed under two headings: direct and indirect.

The direct effect of changes can be measured by identifying how the physical deliverables on a project have been affected and then by assessing the cost and time changes associated with those effects. Clearly, such an assessment does not include all of the effects on cost and time.

The direct effect of a change is the effect that can be associated with the elements of the contract that were directly affected by the change.

Indirect effects can be more significant than direct ones in some instances. These indirect effects include all of those things that are the result of the ripple effect of a change. Some were listed earlier in Chapter 9. We will look at these again later when we consider how these costs are recovered.

The indirect effect is the effect that results from ripple effects, secondary effects and the general disruption to delivery plans created by a change.

9.4.2 Time Effect on a Change

We already seen, with a couple of examples, how the impact of a change will normally increase as we delay making the change. This is particularly important, as it seems that this is not obvious to some of our colleagues. When we ask a contractor to price a change, we should also ask how long this price is valid for. Alternatively, we may ask the contractor to quote a price that is valid for a specified time, as ongoing work may have to be undone or may make the proposed change harder to implement. For example, if we want to add a large piece of equipment, and the building cladding is being installed, we may have to demolish some of the cladding to gain access to where this new equipment is to be installed if you wait more than a week (say).

The impact of a change varies over time.

The very nature of most contracts is that we do work or supply materials. As we complete the required work, we reduce the degrees of freedom available to make changes. In other words, the more work that is done, the fewer options there are to affect work that has not been done. Also, as time passes, lead times become more significant – and harder to accommodate. Finally, we are increasingly likely to affect work that has already been done as time goes by. Clearly, then, time is our enemy when it comes to making changes. This, traditionally, has not stopped people from making changes, but it has led to misunderstandings about the likely impact they may have.

Typically, the impact of a change increases over time.

9.4.3 Contracts

Once in a while we come across a request for a change that is so fundamental that it affects the contract as a whole. For example, a company has been retained to build a software support system that is based on one operating system that they are intimately familiar with and they need to tie into a proprietary accounting system with key information. Then the client changes the operating system and the required programming language. And while they are about it, the accounting system is also scrapped and replaced. In this situation we have taken the contractor out of its area of competence, and we may have an invalid or unenforceable contract.

Consider a contract being awarded to build a high-rise building that has a structural steel frame and a glass curtain wall. There is a change in the ownership of the client organization, and a new architect and a new structural engineer are appointed. The new company is in the cement business, and they want to showcase their product in this building, so they change the design to be a reinforced concrete building with precast panels for the skin. Does this change go to the root of the original Contract and invalidate it? This may depend on what the expertise of the contractor is and on the options that are built into the contract for allowable changes, delays for redesign and more.

Although possible, a change rarely fundamentally changes the contract, although it may result in a significant revision or even termination of the original contract.

9.4.4 Relationships

Can you manage a contract well without good relationships between the principals involved? Is it easier or harder to manage a business deal if the parties know and trust each other? How does the approach to a contractor or a client change if you suspect they have a hidden agenda? Or if there is a grudge from a past experience? Or a bias against the type of organization/culture/product/fill-in-the-blank? For most of us, the answer is that we need to have a relationship with the other party to a contract. Once we remove that relationship, we end up making positional arguments rather than trying to solve the problem. When it comes to a change order that, for any reason, may affect the relationship, we need to step carefully.

It is not uncommon for mistakes to be made or items to be missed in the scope or specifications associated with a contract. When these mistakes surface, they need to be addressed, and the person who is responsible for the preparation of the faulty part of the contract documentation will likely feel somewhat exposed. Understanding the time and other pressures that all of the participants face in the process of preparing such documentation is important. It is usually a buyer's decision about how much time or money it is willing to spend on design that dictates the final quality of the product from that process. Fast or cheap will yield more errors than a more careful (and therefore longer and more expensive) process. Anyhow, we never know when we may need a favor returned in exchange for simply understanding normal and honest human error. Consideration and care in the management of the more delicate changes will help preserve relationships for the long haul.

In the end, the efficacy of a contract is dependent on the relationships between the people involved. If a change is likely to adversely affect such relationships, then the change will require special care in its administration.

Here's an observation, based on a few decades of watching the same problems repeat themselves: Many of the sensitive changes we see on projects are the result of a missed communication of some sort. Consider some of the perennial problems:

- The mechanical specifications refer to the wiring of a switch and interlock for a valve and say it is to be done "by others". The electrical specifications have an identical statement. Who does this work?
- There is an item of excavation for footings and backfill, but none for removal of spoil. Who does this work?
- Who cuts the holes in the ceiling tile for the sprinkler heads? And if it is one trade, and the location is changed by the fire marshal, who pays for replacing the tile?
- I think I'll stop now, or we'll be here all day.

You get the point. And this is the tip of the iceberg. There are changes we need to make to fix problems, but someone is reluctant to carry the news to the customer or to another person who has to make a decision or spend more money. So we wait till the last minute. Our judgment of "the last minute" is not the same as that of the person who is affected and who needs more time to get additional approval or prepare a boss for bad news. These problems are generally solved by working on our relationships and by being sensitive to others' positions and what they need in order to be successful.

Sometimes the simplest issue can interfere with relationships. Commonly this has to do with timely and sensitive communication.

9.4.5 Morale

Heard of the "Russian stop"? This is when we rush to get something done because of an imposed deadline. We work exceptionally hard. We get the work done – and done well. Then the whole project grinds to a stop. Rush and stop. Will we rush the next time the same person asks us to do so? One thing that is worse than this is that, after the work is done, someone asks for the work to be undone. These events are demoralizing. Change orders can lead to such events for the people doing the work. The cost of this demoralization is not easy to measure and explain, but we see it in the changes in attitude of our suppliers and their staff.

Changes that nullify hard work or special effort or that stop work after extra energy has gone into meeting a deadline will demoralize the project team. These and other causes of change need to be managed with care.

9.5 Pricing Change Orders

I am going to preach. Sorry. The sermon is simple and short: *Equity in managing changes and change orders helps us manage them more effectively.*

Why the sermon? Because we seem to play games all too often. Buyers believe that contractors make a lot of money on change orders. Contractors believe they struggle to break even on them. Probably both are right. If a contractor loses money on one change, he or she will try to make up the difference on another.

The calculator is just for show.
He's pricing a change order for more money.

Contractors should not have to lose money on a change order. Equally, changes should not be opportunities to make windfall profits!

9.5.1 What Is Affected?

Earlier in Chapter 9 we saw that a change affects several aspects of a contractor's work. We have also seen that some of the things that are affected are hard to measure. Furthermore, we have seen that there are direct and indirect effects and associated costs and schedule disruptions. All of these need to be addressed in pricing a change order. There are basically two ways of dealing with this. One is to recognize that there are other factors and provide a contingency to deal with the associated risks. The other is to leave the assessment of the effect of this and other changes until later and submit a claim for the losses or other consequences at that later date. Neither choice is really very satisfactory to buyers, especially if they have a need to bring an auditable closure to the change order as quickly as possible.

A change affects the schedule, morale of the participants, sequencing of work, availability of key resources for specific tasks and more. The ripple effect may not be fully measurable in the first instance.

If the effect of a change is not immediately measurable, then the contractor may request that the unmeasured part be assessed later.

Few of the contracts that I have seen set out any rules for determining how "ripple effect" and other secondary costs associated with one or more changes will be assessed. This creates uncertainty and results, often, in loss of trust and erosion of goodwill. It pays both parties to manage the impact of change in an effective way. We know from experience that establishing a realistic baseline for the base contract in terms of schedule, budget and performance is hard to do. We have tried using critical path method schedules to set the baseline for timelines and bills of quantities or unit rates for adding or deleting items so we have a basis for determining the impact on costs. We have techniques for defining scope such as the Work Breakdown Structure and its derivatives and alternatives. Yet reality seems to elude the net we cast. This is because any baseline we set will probably be too rigid a model to accommodate the likely shifts in circumstances that are beyond the control of either or both parties to the contract.

An alternative approach is to determine the likely things that will be affected by the changes we have experienced in the past. Against this list, we can establish percentage markups, unit rates or other units of measure that reasonably represent the anticipated cost or delay associated with the element that is affected. When a change comes along, we look at the factors that come into play and adjust the contract against the schedule we have preset. If the project extends over a longer period, then the schedule may need adjustment from time to time to reflect market conditions. Another option is to preset points in the execution of the contract when a negotiated adjustment for changes to date will be made. We then operate from a clean slate after each adjustment.

A buyer would normally find a potential future cost associated with a change to be disturbing. So it makes sense to have some rules around how the final cost and other effects of ongoing changes may be assessed.

9.5.2 Opportunities and Risks

If we really liked changes in our contracts, we would plan to have more. Generally we do not like to have to issue changes, but we do value the capability to do so because we know that, particularly on large, complex, innovative or urgent projects, we will need to make midcourse adjustments. Each change will inherently carry risk for the contractor and may also involve risk for the buyer. The risks associated with changes include not understanding all of the effects resulting from the change, the need for rework, rescheduling of subcontractors, disruption to the planned operation and sequencing of work, availability of specialists on a revised schedule, materials and equipment delivery if these are not readily available and much more. Not all of the risks will – or can, necessarily – be identified at the time the change is priced.

Another practical issue, from a buyer's perspective, is the need to proceed with a change before the details of the cost and other effects are determined. Often, because of the wording of the contract, this may also be a risk for the contractor. Whenever there is a risk to one or the other party, at least one opportunity exists: This is the opportunity for the party that is not at risk to try to help the one that is exposed. This, however requires a degree of trust.

Just as every cloud has a silver lining, so some of the risk associated with a change may be turned into an opportunity.

There is clearly no obligation to help if not required under the terms of the contract. However, adhering strictly to the letter of the contract as opposed to the spirit of the agreement has some disadvantages, as well as the obvious advantage of being "right." Your position of a contract administrator or advisor is unassailable if the letter of the contract is strictly applied in any situation. However, the give-and-take of normal business is lost, and with it is lost the opportunity to build a sustainable relationship.

Both the buyer and the vendor need to consider the long-term relationship they want to have when working out how opportunistic or bureaucratic they plan to be in the management of any changes. The more important the relationship is in the long term, the more flexible and open the parties need to be.

9.6 Contract Provisions for Markups and what they may Mean

Many contracts try to manage the change process by trying to control as many variables as possible. To this end we often see the percentage markup for a contractor on changes defined in the clauses that address how changes will be managed. Typically, we see 5%, 10% or 15% specified for the permitted markup by the contractor for overhead and profit. Profit and overhead rates may also be specified differently for work done directly by the contractor and work done by a subcontractor.

There are commonly predetermined categories of markups for indirect costs associated with change orders.

Here are a few interesting points about these markups. Many of these points arose out of a study by Cheryl Semple (1994), whose Master's thesis was aimed at understanding the business behind how these numbers were determined and how the changes were managed.

- It seems that nobody knows where the percentages for markups come from.
- There was no agreement in the study regarding what the different markups covered. Here are some examples:
  - Is a working foreman considered to be a direct cost, an indirect cost or part of the cost covered by overheads?
  - Exactly what does the indirect cost markup cover? Small tools? Consumables?
  - Are drill bits and saw blades consumables or tools?
  - Are site costs part of the contractor's overhead?
  - On a $100 change order that takes 5 hours for the contractor to process, is $1 per hour a reasonable rate to pay for the staff member doing the processing (including heat, light, payroll burdens etc) if the permitted overhead markup is just 5%?
  - Equally, on a $1 million change on the same contract that takes the same time to process, is the rate of $10,000 per hour realistic?
  - Do you calculate profit as a percentage on the subtotal of costs before or after adding the permitted overhead?

The list of unknowns and inconsistencies grew to the point that we abandoned the quest to understand the industry standard for administering change orders. We had made an important discovery: There is no standard. Worse, the contracts we looked at in that study, and many that we have looked at since, still do not try to clarify these issues. We believe that this may even be deliberate. I suspect this is because there needs to be some flexibility to negotiate and manage contracts in a pragmatic way. The industry may have found one mechanism buried in the vagaries of change order management.

The markup categories are often ill defined or poorly understood by practitioners and can, in themselves, lead to further grounds for dispute.

Another interesting phenomenon regarding these markups is that they are set by the buyer, not the vendor. It has always puzzled me that buyers know their vendors' business methods and models better than the vendors do. Have we not considered asking the vendors what their markups would be? We can then incorporate this in the overall price and performance assessment that is part of the contractor selection process.

The amount that a contractor may add to cover overhead and profit is often – by some amazing alchemy – specified by the buyer.

9.7   Gamesmanship in Pricing, Negotiating and Managing Changes

We know that all contractors were born at night. The trouble is that it probably was not last night! Most contractors that stay in business will have developed skills that allow them to play the change order game and survive.

If the seller's markups on change orders are specified by the buyer, it is not unreasonable to expect any shortfall to be recovered by an intelligent vendor.

9.7.1 Pricing Change Orders

Let us start with the obvious. Under normal circumstances, assuming there are professional people on both sides of a contract, the best way to negotiate a change is to base everything on as honest and open principles as possible. There may be some practical business reasons for limitations to this approach because, for example, we may need to protect proprietary information or methods. Or we may not wish to disclose specific costs –

again, this tends to be because of concerns over access to this information by the competition.

The best way to negotiate changes is to do so based on honest and open communication, as long as this may be commercially reasonable.

A contractor should be able to recover all of its costs associated with a change. This would, of necessity, need to include covering risks, indirect costs, impact costs resulting from changes to the sequence of work, productivity losses and other reasonable charges.

The pricing of a Change Order should reflect not just the direct cost but also the cost and time associated with all of the effects of the change on other work.

Some industries carry scars from consistently bad practices. Many construction buyers, for example, have had to deal with what they believe to be unfair practices by contractors. They worry about price gouging; games are played on such a regular basis that any contractor or buyer, consultant or contracts specialist will have a marvelous collection of stories. Because of this history and the many cautionary tales from all perspectives, we have built a set of approaches to the management of changes that are aimed at minimizing exposure to future problems. Perhaps one of the most common of these is requesting unit prices for specific things that are likely to change after award of the contract.

It is common for buyers of construction goods and services to request prices for specific components that they believe may change. These unit prices may be written into the contract to be applied to subsequent changes to the contract.

It is also common to include in unit rates of this kind an "add" price and a "delete" price. The reason for this is that there may be restocking charges, costs associated with the transportation of materials and other hidden items to account for in a deletion. In an addition, there will be supervision charges and costs associated with small tools, consumable materials and other incidentals that need to be added to the base price.

Typically there will be a different rate for adding quantities versus reducing them.

Most contractors prefer to submit prices for changes that reflect commercial reality as they see it rather than to submit the cost plus a predetermined markup to reflect overhead and profit.

These examples are based on a stipulated price contract. Other more flexible contract forms such as unit price or cost plus contracts will have different change mechanism built right into the payment method. As discussed in Chapter 2, a cost plus contract is one in which the contractor is paid for all of its direct and indirect costs plus a fee for services. The fee is commonly a fixed fee, but occasionally this is based on a percentage of the actual cost of the work. In addition, there may be some performance incentive. Adjustments to the fees and the incentives may be required under certain types of changes that significantly alter the contract.

In the case of unit rates, the contractor is paid for the number of units installed or otherwise worked on. If you change the quantities, so you change the amount paid to the contractor. The problem is that we may need to change the unit if, for example, the original unit rate was for excavating topsoil and sand, which turns out to be clay or rock (which is either heavier or has a larger bulking factor, so more truck trips are needed to dispose of it, etc.)

Also, in the case of unit rate contracts, estimated quantities are commonly provided. If the quantities are modified significantly, then the rates may be affected. Some contracts allow for adjustment to unit rates if quantities change by a significant amount from those that were anticipated.

Even unit rate contracts may allow for changes in the unit rates if the estimated quantities at the time of bid differ from actual ones by a given amount (often 15%).

9.7.2 Negotiating Change Orders

Negotiating a change order price is part of what we need to do if we want to be comfortable that the price is fair and reasonable. This too can quickly become a game (see Only One-Third in Section 3.4.3).

There is no need to negotiate a change order that has been fairly priced and presented to a knowledgeable buyer.

If either party to the contract feels a need to negotiate the price of a change order, it pays to understand why this need exists. If we can understand the anxiety of the concerned party, we can try to make adjustments to how we work with the change order process to make it more transparent.

If a change order needs to be negotiated, then check why and adjust both pricing practices and management of that change order accordingly.

Contractors who try to make a windfall of money through change orders do so because they are greedy, because they are losing money on the rest of the work or because they simply enjoy the game. Whatever the reason, we must be prepared to have our pricing quizzed and challenged by the buyer or his or her consultants.

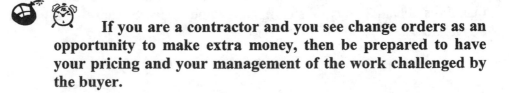
If you are a contractor and you see change orders as an opportunity to make extra money, then be prepared to have your pricing and your management of the work challenged by the buyer.

The converse also holds true. If you are a buyer and you are always challenging every price quoted by the contractor, be prepared to have the contractor play games with you as he tries to ensure that he can recover his costs and a reasonable profit in spite of your vigilance!

 If you are a buyer and you believe that the contractor is trying to make more money than it should on change orders then take the time to understand why the contractor has taken the approach it proposed and question or challenge any perceived or real inappropriate practices.

9.7.3 Managing Change Orders

Whether we are a buyer or a seller, if we are trying to act professionally, we will need to be objective in our assessment of the effects of a change order on cost and time. We must also consider the effects on project performance, safety, morale and quality. The effect of the change order should be fairly reflected in how it is priced. For many contractors, cash flow is critical, and if the change disrupts this cash flow in a negative way, there may be additional costs arising out of the need to extend a line of credit or otherwise finance the work until payment is made.

Equally, if the buyer is financing the project through revenue-based cash flow, then a temporary "bump" could require liquidating other assets or seeking temporary financing. This may delay payments to the contractor.

Make sure that the effect of a change order is reflected in both the final cost and in the cash flow requirements for the contract.

Between developing the specifications for the change order, getting it priced, accepting the cost, time and other effects, then implementing the change, we have all of the ingredients of a minicontract within the main contract. This is perhaps not a bad way to think of changes. If the change is not processed for any of a number of reasons, it may be because we have a dispute. Chapter 10 looks at disputes on projects and how they may be managed with minimum disruption.

Managing a Change Order is like managing a minicontract within the main contract.

Chapter 10

THOU SHALT NOT KILL
(YOUR CONTRACTING PARTNER)

> *Claims and Disputes:*
> *Resolution Options and Avoidance.*

If changes are inevitable, then claims and disputes must be a close follower. Chapter 10 discusses how serious a claim or dispute becomes and when it crosses the Continental Divide from resolvable to irresolvable. Here, again, work done by the Construction Industry Institute (CII) sets the foundation for this discussion. The report "The Continental Divide" produced by the CII in 1995 identified a point at which a contract dispute went from relatively easily resolution to difficult resolution. Dispute resolution happens most easily when it can be negotiated. Other forms of dispute resolution are also worth considering.

Chapter 10 looks at the range of dispute resolution options available, from negotiation through to litigation. We will then look at my favorite form of dispute resolution: not having a dispute in the first place!

 Although Chapter 10 looks at dispute resolution methods, let us not lose sight of the fact that the best form of resolution is not to have the dispute in the first place.

10.1 Negotiating Changes and Disputes

Like many others who have been in business, I have had to negotiate or renegotiate a deal. The process of negotiation is complex, and explanation of the skills required goes beyond the scope of this book. Negotiation, however, is something we inevitably do at some point in the execution of anything but the simplest contracts. We may negotiate before the contract is entered into, and we will certainly do so afterward. We will negotiate changes, revised terms and conditions, extras and more. When negotiating, it pays to try to start the process before people have become entrenched in their positions. Once this has happened, there is more at stake than just the issues. We also have to deal with politics, the positions themselves, and the perceived sense that each side has to win.

It is always best to negotiate before people take positions.

If we think about what negotiation on a contract is, we really are trying to establish a basis on which we are to proceed with the contract. If the negotiations lead to some change to the contract, then we issue a change order. If they result in agreement that the contract already addresses the issue, then we have resolved the issue and can continue without a formal change.

Chapter 9 looked at changes and how to manage them. Negotiating a dispute really means turning the dispute into a change.

We can learn something from this. If we have a change, it is something both parties agree is the result of doing something that was not contemplated by the Contract. Thus, we can argue that a dispute is something that one party to the Contract feels is covered by the contractual agreement and the other party feels is different to the intent of the Contract. The lesson is that we need to encourage simplicity and clarity in the contract documents, and work toward alignment of the parties. If we fail, as we so often do at the outset, because of the rush to get going, we need to work at this alignment as we come across the problems. These problems generally emerge in the guise of a disagreement as to what the Contract was intended to mean.

A key difference between a dispute and a change is that a change is a modification to the Contract that both parties agree to. This requires finding common ground.

When we come across a dispute, we really need to determine if there is some entitlement to compensation. If there is (usually based on an *objective* interpretation of the Contract), then we can ask ourselves the next two questions. These are, in order: What is the claim really worth? And how much are we going to ask for?

If we submit a claim, it should be based on entitlement under the terms of the Contract. This implies that we have read and understood the Contract before entering into it. It also means that we understand the Contract and have confidence in our position. Then we have to work out what the change has done to our ability to make the expected profit on this project. We need to make good the losses that result from this change and we need to be able to explain how we arrived at the final number. Normally, a contractor is entitled to a reasonable profit and recovery of reasonable overhead charges. As we saw in Chapter 9, the reasonableness of these markups is often determined by the buyer and included in the terms of the contract.

The final step is preparing to negotiate. Although I prefer a simple and honest approach, we do not always seem to have this luxury. In situations in which there is little trust and the buyer or the seller is trying to play games, claims are often "fluffed up" a bit, so there is something to give away.

In negotiating a change – and a dispute becomes a change if successfully negotiated – we need to have answers to some key questions.

If we have exaggerated our claim in order to give something back, we already understand the concept of win/win – even if this version is a bit dubious. Even in a straightforward and honest negotiation, there are things that we can negotiate without giving up what is important to us. We need to listen carefully to the other party to understand their position. There are probably two important parts to this.

The first part happens before the change or dispute arises. At this point, it pays to understand if the client has the capacity to pay for changes. If the buyer is stretched financially, then recovery of anything beyond the Contract, and sometimes even entitlement under the Contract, is going to be hard. Some due diligence at the outset pays. Consider, when appropriate, looking for cost savings in other parts of the project to offset the additional cost of the change.

After a change, it pays to take the time to help the buyer's representative understand where the exchange of value is. In other words, if we are asking for payment, it helps the person we are asking if we provide the convincing argument that they need for their bosses, shareholders or auditors. Equally, if we are the buyer, it is useful to have a real understanding of the costs we are being asked to pay for. Without this knowledge – which we may wish to verify – it is hard to authorize and then justify an additional payment. In many cases, an experienced contract administrator will know why things cost what they do. This does not mean we need to make the assumption that they do. It is this type of exchange during the negotiation that leads to win/win situations. It is important for sustainable relationships that we have win/win solutions whenever possible. Failing that, at least work toward a lose-as-little-as-possible/lose-as-little-as-possible solution.

If we have had a successful negotiation, we and our negotiating partner will feel that we have each been successful. Invariably this means that we have both compromised.

Sometimes we need to give up more than we feel is fair. We need to take a step backward and look at the big picture, and then readjust our position. I recall one of our favorite clients once asking in the middle of a discussion whether, overall, we made money on their projects. We had done work for them for many years, and their account represented as much as 25% of our work some years. The question came out of the blue while we were working out a price for a change order. We were a long way apart in our views, and I could not fathom why, because we usually were quite close. Because the question was such a sudden change in direction in the conversation, my brain switched on. The client needed a favor. Could we cut them some slack? Maybe they had made a bad decision and the change was needed to fix the problem. This clearly was not a topic for discussion. And they were a good client. And we made a good margin on the work we did for them. Giving up a bit on this occasion (a loss on the change order, lost profits on the project) was a relatively small deal in the larger scheme. So we gave up on our margins, absorbed a large part of the cost on this change and built a stronger relationship with our favorite client.

Never underestimate the value of a sustained relationship. It is the path to more business. Presumably we want that business only if it is profitable in the long run.

10.2 Mediation

Sometimes we simply cannot see the wood for the trees. Negotiation is something we all believe we are good at. These two factors blind us to the need for outside help when we really need it. When we get stuck in negotiation, it is often useful to get a neutral third party involved in the process. They can mediate. This is of particular value if we have reached an impasse or have taken positions that we are not willing to move from. Good mediators have three assets that help in managing negotiations that have failed or are about to fail. First they are not involved. Second, they are the right type of person (sufficiently patient, good listeners, creative etc.) to help us find possible solutions or to encourage us to be more innovative. And finally, they are trained in the skills of mediation. Based on my experience, the very best mediators are half way between facilitators and arbitrators. This comes with experience, as they will have seen many disputes and will have helped people come to a resolution. There is a good chance that they can guess at what we would likely do in a given situation.

Mediation is half way between negotiation and arbitration.

One of the key differences between mediation and negotiation is the presence of a third party. This helps us behave more professionally. It helps us listen to the real issues and address them rather than staying in the corners we have painted ourselves into. In the end, we will have found our own solution. It is easier to own something we have participated in. Entering into mediation, and doing so with a real intent to come out of the process with resolution to our problem, requires a commitment to this outcome and in turn requires acceptance of the role of the mediator.

Mediation recognizes that a third party will help the negotiating parties find a solution.

How do we set about a mediation? We need to have agreed to the process first. Then we need to select a mediator. The "rules" for the mediation also need to be agreed upon. Finally, and before accepting this option, we should understand what mediation can and cannot do for us.

What are we agreeing to with mediation? Simply put, we are agreeing to have a third party work for <u>both</u> disputants with the intent to come to a resolution. Mediation is nonbinding, and mediators have no authority to make decisions on behalf of either party. If mediation is desired, check with professional and technical associations, which often have a current list of trained mediators to choose from. Business organizations and Chambers of Commerce are other potential resources. Also, lawyers can often offer referrals to mediators, because, increasingly, the courts ask disputants to go through mediation before they will hear a case. Both parties need to agree on whom the mediator will be. Typically this means that both or neither party knows the mediator. In the former case, both parties should feel comfortable with her neutrality. In the latter case, they should interview candidates and agree on one.

 Mediation is nonbinding.

The process for the mediation should be agreed on ahead of time, and decision-makers for both parties should be involved throughout. Good mediators offer suggestions including location, premeetings and a description of what to expect. The premeetings help the mediator understand both perspectives. Through this, the mediator can point out the realities (strengths and weaknesses) of both positions and help generate possible solutions.

Mediators bring process that is based on experience. If they are trained and certified mediators, they also bring experience and knowledge derived from this training. In addition to navigating through the process, the right mediator will help in other ways. Many experienced mediators have knowledge and understanding of the issues at stake. Even new and freshly trained mediators have knowledge that they bring and that is different to that of the principals in the dispute. All of this adds to the richness of the renewed negotiation.

 The mediator often provides expertise and process to enable discussion and resolution of a dispute.

Because we are human, many of us see mediation as the result of our own failure to negotiate. We rarely turn to this option in resolving a dispute. When mediation has been used, it has reportedly had a remarkably high success rate. The failures – based on a personal small sample – have been when one party pays only lip service to the process.

 Mediation is often a missed step in the resolution of a dispute. Given a chance, it has an inordinately high success rate.

So, mediation may not be the solution in some cases. We have a hard nut to crack. The next step in escalating the dispute resolution process is to resort to a third party who will make a decision for us. In this case, we may pick the third party, and at least retain some control over whom the "judge" will be. We can also elect to go to either binding or nonbinding arbitration.

If mediation is clearly not going anywhere, then we need someone to come up with a solution on our behalf. This is when arbitration may make sense.

10.3 Arbitration

Many North American legal jurisdictions have formalized or legislated the arbitration process. As a result of this, it has become a scaled-down version of litigation in some instances. The distinction is still important, as the proceedings remain private and there is some choice over whom the arbitrator or the board of arbitrators will be. Also, the cost is generally lower and the time for the process is faster than for litigation. One of the critical differences between mediation and arbitration is that the decision is no longer made by the parties involved, but by the arbitrator.

Arbitration is a scaled-down version of litigation in that others make a decision based on our evidence and that of our opponent.

Arbitrators are selected in a number of ways. One of the most common is to select from a list agreed to by both parties. Another common approach is to pre-name the arbitrator in the contract document. A third approach is to choose from a select panel such as the American Arbitration Association. Yet another way, used often too, is to work with a panel of arbitrators. In such a case, it is common for each party to select one of the panel members and then for the two appointees to select the third person. Either way, there is a chance of influencing who the judge may be. The obvious advantage of this approach is that the decision-maker may be someone familiar with the practices and business of the industry in which the dispute occurred.

In arbitration, the "judge" often has knowledge of the business we are in. If a panel of arbitrators is used, we will have a say in the selection of the panel.

Another area of choice that is open at the time of tender (or afterward if not specified in the contract documents) is the decision to use binding arbitration versus the nonbinding variety. What is the difference? In binding arbitration, it will take a court (as in an appeal against a court decision) to overturn the decision of the arbitrator. This is generally the case, although regional differences in the statutes that may govern this process can vary. In nonbinding arbitration, it is common to have a requirement stating that the decision of the arbitrator remain in place until the end of the contract, at which time any remaining or outstanding differences may be revisited.

Arbitration can be binding on the parties that enter into the process.

 It is common to include an arbitration clause in a contract. Typically this will be for binding arbitration as an alternative, in the first instance, to litigation.

The "grand contract litigation" starts the week after.

Arbitration, as an option, needs careful consideration. Even if it is not mandated in the contract, it remains an option at any time, provided both parties agree to the process.

Arbitration is always an option, even if not stipulated in the Contract.

The arbitration process typically follows rules similar to those of a court of law. There is a greater flexibility, in most jurisdictions, regarding rules for evidence, and often there is greater flexibility in the judgment of the arbitrator, as there is usually no need to worry just about what is legally correct, but also what is fair and reasonable.

A key difference between arbitration and litigation is that we can include the idea of fairness.

These points demonstrate how arbitration is an option to consider if litigation is to be avoided. The next option, then, is to litigate. This is expensive and time consuming. The outcomes are also far from certain.

10.4 Litigation (Classic Dispute Resolution)

Before embarking on a brief discussion of litigation, I must confess to another bias. This stems from observation, as well as from experience. I believe that litigation is a sign of failure. We are in court – with the attendant stress and problems – only if we have failed to resolve a problem like grown-ups. Unfortunately, there are some organizations that give us little choice. These are commonly the ones that pass any dispute on to their lawyers for resolution. One of the indicators that I watch for is when corporate counsel is the go-between. I would like to avoid working with any organization that does this, because in my business I do not normally deal with legal issues, but with the business itself. If I need to negotiate through another party's lawyer, I get the message that I am too dangerous to do business with or that the other side is not interested enough to get to know me. There is no trust and no opportunity to build trust. I am better off walking away from such a relationship.

Here is another personal bias: I believe that we have failed in some way if we end up in litigation.

Litigation, to me, is like a cow being fought over by two farmers. One is pulling the horns, while the other is pulling the tail. In between is the lawyer: milking the cow. This is not intended as a slur on the noble profession of law but as a comment on human behavior in general. So, please do not quote this analogy out of context! Lawyers manage only the problems we bring to them. Mistrust is what leads to much of the work they do. Litigation is expensive. I have only entered into it in total desperation, and with the expectation of, at best, a 50% chance of winning – and that is based on the assumption that I believe I am totally correct in the position I have taken.

If we enter into litigation, we should be prepared to lose. We have a choice: Lose by a little, or lose by a lot.

Let us take a look at the steps involved in litigation. What follows is generic, as the detail varies depending on the legal jurisdiction we are working in. Table 10-1 shows the general steps and a range of times that this step may take. I cannot even hazard a guess as to the cost, but a study by a law firm in Canada suggested that it tends to exceed the amount at stake in any event.

Litigation Step	What Happens	Range of time this step may take
Brief legal counsel	We tell the story as we see it and get legal advice. The legal team will review the documentation and evidence and determine the value of the claim.	1 month to 1 year
Prepare claim and file with the courts	Legal counsel prepares the claim and refers to us and to expert witnesses. The claim is filed with the courts, and the defendant is served appropriate notice.	3 months to more than 1 year

Litigation Step	What Happens	Range of time this step may take
Defendant response	The defendant responds to the claim in one of several ways. Settlement may be possible. If settlement does not occur, we continue to the next step.	1 to 6 months
Examination for discovery	This process requires each party to disclose its evidence to the other. Part of the purpose of this is to encourage settlement of the dispute and to see if there is, in fact, as strong a case as we first thought. Although the opportunity clearly exists to see the other party's perspective, this tends to be a more aggressive process, searching for weak spots in the other side's case. Settlement may be possible. If settlement does not occur, we continue to the next step.	3 months to 2 years
Pretrial and date set for trial	The pretrial, if it takes place, usually involves both parties, their advisors, including legal counsel and a trial judge, although not the one that will hear the case. The idea, once again, is to see if a resolution can be reached. If no resolution is reached, a trial date is set. This can be years in the future.	Including organizing the event and scheduling it, this takes at least 3 months and can take much longer.
Trial	The trial itself will take days at the least (unless it is in a small claims court) and may take weeks or months depending on the money at stake and the legal issues and precedents that this case may set.	Including the waiting time for an available court and judge, this can take years.

Litigation Step	What Happens	Range of time this step may take
Decision of the court	This is the time required for consideration by the judge – or if this is a trial by jury – then the time required for the jury to make a decision.	In Canada, where trial by jury on civil suits is not an option, trial judges have taken more than 1 year to bring down a decision.
Settlement or appeal	The game is not over yet. We have a judgment but do not have the money. This may still be negotiated between the parties. Also, an appeal to higher courts may be an option for the loser. In this case, we are many years from any settlement.	This step can take days (very lucky) to months or even years if there is an appeal.
How long altogether	Assume there is no appeal.	At BEST – more than a year; at WORST – 5 to 10 years or more

Table 10 -1. The litigation process time frame.

The litigation process is extremely costly and time consuming. We need to remember that the process is a distraction from our normal business and requires management and other attention at times that are out of sync with the real business issues that drive our operations and projects.

Litigation is a formal, expensive and costly process.

The litigation process is also a public one. This means that anything that appears as evidence in the court will be available in the public domain. As the legal arena is an unfriendly environment, we can also be sure that our dirty laundry will be displayed in its worst light. Even some of the clean laundry may look dirty after it has been presented in court.

Be prepared to hang your dirty laundry out in public.

The final point is really important. Litigation is about applying the law to a problem and getting a legally accurate resolution. It is not about fairness. Fairness will almost certainly be left behind when we abandon negotiation and mediation. Depending on whom the arbitrator is, we may have left fairness behind when we went to binding arbitration too!

 Litigation is NOT about being fair – it is about being legally entitled. There is a BIG difference.

If litigation really is the right answer for a dispute, consider doing it faster through a privatized form. This is not available in all jurisdictions, but if it is, the private system is often referred to as a mini-trial.

 If we are determined to use the litigation option, a possibly faster route in some jurisdictions is the mini-trial.

10.5 Mini-Trials

We won't spend a lot of time on this section, but it is useful to be aware of the options in litigation. In some jurisdictions, the courts are so backlogged that other ways of bringing issues to trial are sought. Privatizing the process is one option that has emerged. The mini-trial uses private halls or rooms instead of a designated courthouse. The judge is either a retired judge or a leading trial lawyer respected by both parties. The rest of the process is identical to the regular court system, except that the waiting line is shorter, so the matter can come to trial faster. When there are statutes in place governing this type of trial, the results are just as binding as if they were the outcome of a regular trial in a designated court of the state or province.

 A mini-trial is available in some jurisdictions. Basically it is a privatized version of litigation.

Mini-trials are usually governed by all of the rules that apply to litigation. They can be more expensive but should be faster.

Fortunately there are other ways of addressing the problem. These carry a three-letter acronym, indicating that they are credible and have been around long enough to get into the jargon of contracts specialists. The three-letter acronym is ADR: Alternative Dispute Resolution.

10.6 Dispute Resolution Boards

ADR techniques are growing, as people try to find better ways to address issues and problems. One specific form of ADR is the use of a Dispute Resolution Board. The idea behind such a board is that it is used to resolve problems as they arise on a project, as an alternative to the more costly option of litigation. The basic concept has been around for a long time. And it has been updated and modified from time to time.

 Dispute resolution boards have been around for a long time. They have been refreshed from time to time.

How does this work? The principle is that a board of adjudicators (usually three) make a decision on a dispute each time one is presented to them. There is usually a process that escalates the problem to the board. This typically takes the form of time limits on resolution at different levels in the project before it gets nudged up to the next level. The board may make binding decisions, although it is much more common for the board's decisions to be binding only until the project is complete. Once the work is done (the board having made decisions that at least permit the business and technical issues to be set aside temporarily), the parties resolve any outstanding differences. This may mean revisiting any or all of the decisions of the board. Success is really obtained by encouraging resolution before the board gets involved. This happens almost automatically, as people are averse to getting third parties to make decisions on their behalf.

Members of a dispute resolution board need to be picked with great care, as their credibility will affect their acceptance and therefore their success. A good choice for this role would be qualified mediators. Consider the mediator selection options in Section 10.2. A "one-person board" is often referred to as a "Standing Neutral". This latter option often serves well.

We need to keep the processes up to date, fresh and relevant if they are to have sustainable success.

Consider also a "Standing Neutral," who is, in effect, a one-person dispute resolution board.

10.7 Other Options for Dispute Resolution

ADR as a term tells us that there are alternative solutions to the more traditional ones. The range of resolution options is huge, simply because we keep trying to do things better. Chapter 10 covers only a few options. There are many variants on these options. I believe there is no right one, but there are better and worse ones depending on the situation and on the attitudes and expectations of the various participants in a dispute.

Mankind is creative. There are many variations of the dispute resolution options described in Chapter 10.

Having said that there is no single best option for solving a dispute, I must now acknowledge that there is an option that is better than all of the others: Do not have a dispute in the first place. This takes more planning, more communication, and more sensitivity to the concerns of others than we normally provide in managing the procurement process.

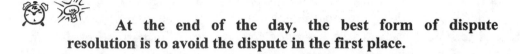

At the end of the day, the best form of dispute resolution is to avoid the dispute in the first place.

If avoidance is the best solution, what do we need to do? The key seems to lie in several areas that are discussed in Chapter 12, where we bring together some of the best practices in procurement and supply chain management that we have observed in industry and that have been tested by dedicated researchers and practitioners. Dr. George Jergeas (1994) has done some wonderful work in turning regular contracts into real partnerships on live projects. His experiences and action research have validated some of the practices that have also been tested in other areas, although perhaps not as rigorously. Chapter 12 provides an overview of how we can build better and more sustainable working relationships that yield greater value for both the buyer and the seller in a contracting relationship.

 If disputes are inherent in any human interaction, then we need to find the format for such interaction that minimizes the incidence of disputes and maximizes our chances of surviving them.

Not having a dispute means careful consideration of how we enter into and manage contracts and, more importantly, how we manage the people who are connected to these contracts. This is the topic of Chapter 12.

Apart from the 'bang' from the demolition charge,
there doesn't seem to be any other impact.

Chapter 11

SMART CONTRACTING: AN INTERLUDE

*The contractor modified the design when
the owner said he was not inclined to pay him.*

I do not think any contract – and therefore any book on contracting – should end up in Chapter 11. So Chapter 11 in this book serves to make sure that neither you nor I end up there! Chapter 12 tells us how to put the 10 commandments of good procurement or contracting into practice in a cohesive way.

 I have fought to keep Chapter 11 short because we need reminding – especially in the business of contracting – that we can take life and all it offers us far too seriously!

 It pays to "lighten up" when working on contracts and relationships. Why not have some fun?

End of Chapter 11.

I did say "heads will roll" and "we face major cuts"
but you are taking these statements out of context.

Chapter 12

SMART CONTRACTING: A FRAMEWORK FOR BETTER PERFORMANCE OF CONTRACTS AND THE PEOPLE INVOLVED

> *It pays to be SMART about how we*
> *manage our clients and our supply chain.*

Chapter 12 pulls together the different elements covered in the other chapters. Each chapter so far has inherent in it a "contracting commandment." Following these commandments in an organized and business-like fashion leads to more effective contracting. In this final chapter, we will look at a framework for implementing these commandments.

This approach is referred to as SMART Contracting™. The acronym "SMART" stands for *S*trategically *M*anaged, *A*ligned, *R*egenerative, *T*ransitional. SMART Contracting is a balanced approach that considers technical, legal, business and human issues in the development of more effective contracting approaches while recognizing the need to accommodate ongoing change. It is deliberately nonprescriptive. This means that we need to apply the principles and manage the detail depending on what we want to achieve and how much latitude we have to adapt corporate procedures.

I feel that I have to apologize for the trademark. In a previous book I used other trademarked terms and was criticized for doing so by one reviewer. I am very sensitive to such criticism, especially if I agree with the comment! However, the trademark protects the term.

SMART Contracting is used to cover a whole approach to supply chain management that has led to improved profitability and competitive edges in field-testing and in application. Although the most basic concepts are covered in this book, most specifically in Chapter 12, the detail would have made the book many times thicker and a lot harder to digest. Any efficiencies relayed reflect but a part of what we call SMART Contracting which, in turn, is a part of SMART Management™. SMART Project Management™ (another part of SMART Management) is outlined in the book "Don't Park Your Brain Outside: A Practical Guide to Improving Shareholder Value with SMART Management," published by the Project Management Institute (2000).

12.1 Thou Shalt Contract within the Law and the Working Environment of the Contracting Parties

> *Contracts are legally binding and other obligations exist beyond just the terms of the contract. These obligations need to be understood and agreed upon ahead of time and in the context of today's business world.*

If we visit the barber's shop or the hairdresser, sit down, and let someone start to cut our hair, we have a legally binding contract in common law. Chapter 1 covered the basics of common law and touched on the law of tort. We know that we do not need a large contract for a haircut. So what *do* we need for an effective and binding contract for something more complicated that we are trying to acquire or sell? The purpose of the contract is to provide a shared picture of what is being exchanged and how it will be done (time, responsibilities, etc.). We recommend a minimalist approach to assembling a contract.

 We already have a basis for a contract and how to administer it in the legal framework we live under.

When renting a car, do we always read that pale grey fine print that covers the back of the agreement we sign? When parking our car and paying for the privilege, do we read the terms and conditions, limits of liability, and other parts of the deal we are entering into? How about that last airline ticket we bought? Did we read the rules that we agreed to be bound by when we purchased the ticket? Even when we have more than just the basic common law contract terms, most of us typically do not bother to really read or understand what those terms are. We rely on commercial recourse rather than legal recourse. If the vendor we are dealing with does not treat us fairly, we will either never use that supplier again or we will tell our friends about the experience – or both. We carry weight in the marketplace. We know it is normally not worth filing a lawsuit, so we find other ways to show our displeasure. Should we be doing something similar in our business dealings? After all, how many of the people involved in delivering on a contract actually have access to the contract document? And of those who do get to see it, who actually reads it? And really understands it?

 Why do we need complex contract documents, when many of the people working on our project do not read them? We are better off relying on commercial recourse than on litigation. Our contract documents can reflect this by being as simple and brief as possible.

If we stop to think about who really reads the contract, we realize that we need to work in as natural a way as possible, as most of the people who have to deliver something under the terms of the agreement will not be exposed to the document at all. As it happens, common law provides a sound and fairly intelligent framework for conducting business. Generally speaking, we cannot contract out of the law. So it pays to mess with the framework as little as possible, as we put our private law together.

There are some basic rules to follow in the business of buying and selling. We cannot contract out of the law.

Our private law is the one we create, and agree to be bound to, when we write and sign a contract. This private law is enforceable in the courts. If our private law is complex, or if it is in conflict with the public law, then our lives will be more difficult, as we have increased the uncertainty associated with the contract's interpretation.

There are two parts to the law: the law that applies to everyone and the private law that we create for ourselves with our contracts. These two parts should be aligned.

What this means is that we should start with the law as it stands and introduce to our contract only those elements that are not precise enough or comprehensive enough to address the specific needs associated with our project and circumstances. There are potentially many of these things to consider. They could include the amount and timing of payments, the technical specifications of what is to be delivered under the contract, rates for use of equipment or charges for people's time. Maybe it is necessary to specify standards for safety, liability for specific risks or actions and so on. It pays to review our existing contracts and standard terms and conditions with a minimalist perspective. There are likely to be redundant or conflicting clauses. We may be asking for protection against something that is unlikely to happen but which we would be protected against, in any event, in common law.

It pays to keep these private laws simple and not in conflict with the public ones. This leads to a minimalist approach.

I am not a lawyer. So when I need legal advice, I go to someone who knows what he or she is talking about. The law differs from one jurisdiction to another, so it is wise to access local expertise as well.

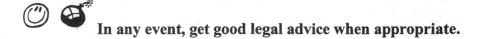 **In any event, get good legal advice when appropriate.**

We also encourage the use of a project charter when appropriate. Any contract is effectively a project. A good project charter serves several purposes: First, it is a living document that is amended by the participants to suit the changing circumstances and needs of the project, its sponsor and, when appropriate, other stakeholders. It is possible to incorporate the charter and a process for amending it in the contract. It can replace just about every part of the Contract, except the Agreement.

 It usually pays to work with a SMART Project Charter™ for each contract we enter into. A SMART Project Charter will help us align the contracting parties by improving communication and setting measures of success that include important issues such as priorities for the project and factors that are critical to the success of the project.

Current practice tells us that we need to have a specification and drawings, supplementary conditions and other elements or their equivalents to a contract. These are sound as they are well understood. They are also perceived as being static and defined. Reality today is not like that. Change is inevitable, so we should consider taking advantage of what we have already established and add only what we need to the contract documents in order to achieve alignment and clarity. As there are other forms of protection in the legal environment we live and work in, we do not need to repeat any of these in the contract, unless we wish to modify or avoid them.

Note, though, that some statutes (e.g., the Lien Act) can often not be contracted out of. A useful tip in determining what we can contract out of and what we cannot is to ask a question. If the answer to the question, "Will contracting out of this piece of legislation affect a third party?" is positive, then we probably cannot avoid it being a part of our deal.

That said, the point of all of this is to highlight that we can establish a really tight and inflexible contract. Or we can take a minimalist approach. In the former case, we will likely get support from all of the really traditional and conservative elements of our organization. They will perceive security and certainty. This is wonderful at the time of entering into a contract. It may remain wonderful to the end if there are no – or few – changes and the outcome is highly predictable. This is true only for low-risk situations. In low-risk situations, the contracting process is usually predictable and should not be changed, as the benefit is low. In higher-risk situations, however, we are fooling ourselves if we believe everything is going to remain static. We are making life more difficult as we appear to imply greater certainty than there really is. This raises the expectation of certainty with just those people that we will later need to go to for approval of additional funding or for accepting changes that are required for the delivery of the product we have contracted for. Recognize the level of uncertainty and reflect this in the form of the contract.

"Watertight" contracts are as leaky as the nearest loophole.

We need to consider the trade off between being right at any cost and being flexible but with some risk.

Clearly, if we work with a more "flexible" contract or purchase order, we are exposed by that very flexibility. This, however, is a more realistic and, therefore, a more manageable risk than the high risk that exists when working with a rigid contract. If the risk is high, there will be a premium in the price we are quoted. If there is no premium, the contractor could lose money and we are likely to have a dispute or bankruptcy on our hands. This can become very expensive in a really bad situation, significantly adding all sorts of costs to the nice "firm price" that we started the contract with.

There is greater opportunity and greater apparent risk in working with a more flexible agreement rather than with a formal and fully detailed contract.

We work in a world that is quite different from what it was 20 years ago. Over the past two decades, most organizations have struggled through total quality management, business process re-engineering, corporate re-organizations, refocusing on core values and products, the partnering and alliances bandwagon (see The Partnering and Alliances Bandwagon), outsourcing, and right-sizing. Many who live in the wake of these are tired of changes. So why change anything now?

Consider just one element of what many organizations have done. We have become more focused on the parts of our business where we add value. As a result, we have outsourced things that we would not have done a decade ago. We may have outsourced warehousing, manufacturing, accounting, training and more. The more we have done of this, the more reliant we become on our suppliers for the very survival of our business.

We are locked in to suppliers that can ruin us if they fail. Think about power (energy) suppliers, Internet suppliers, telephone companies, and so on. If an entire industry stumbles (e.g., airlines in the early 2000s), we start to change the way we do business. What would happen if telephone companies started to go bankrupt at a rapid rate?

The Partnering and Alliances Bandwagon

The terms "partnering" and "alliances" have been bandied around, abused, and misunderstood. They are now viewed with caution by many practitioners. Why? Because too many badly planned and managed partnerships and alliances have left a wake of failed projects, damaged careers, and even lawsuits behind them. Even well-managed partnerships and alliances will fail if they are founded on incorrect assumptions, insufficient budgets or schedules, misconceptions, or inflated and unrealistic expectations.

That said, I have yet to see a well-planned and managed partnership or alliance fail. Successful implementations of such relationship-based contract types require a number of key ingredients. These often include the following:

- A clear and shared mandate from all parties, usually articulated in a project charter
- A standing neutral (see Chapter 10) with the right experience, knowledge and wisdom
- A good and flexible plan for execution of the work
- Ongoing maintenance of relationships often supported by a well-thought out communication plan
- The ability and decision-making power to manage changes and respond flexibly to evolving circumstances.

This is not an exclusive list. The cost of the additional effort required to establish an effective partnership or alliance is typically very small in comparison to savings in transaction costs alone.

(Cont.)

The Partnering and Alliances Bandwagon (cont.)

The effort required for effective partnering is far out-weighed by the benefits.

A significant element of effective partnering is that it can be implemented through a structured process that addresses each of the three areas we need to deal with. We need clarity and alignment of business issues and drivers that require the contractual relationship in the first place. This first requirement is best addressed through an aligned view of what successful completion of the contract and project will achieve. The second requirement is to deliver the technical products. So these need to be developed in such a way that we address the right problem with the best solution. We have been known to try and build a relationship around trying to solve a solution rather than the problem. Finally, we need to address the social issues that can damage our relationships through traditional attitudes.

Partnering that works requires structure and planning.

It is particularly important that a partnering or alliance approach is recognized for what it is: It is essentially a framework on which to hang good practices. The ideas presented in this book have demonstrated success time and again. Lip service, label changing and head nodding will *not* get better results. We need to work with our business and contracting partners to change views, attitudes, habits and, eventually, results – all for the better.

The rest of Chapter 12 provides recommendations and other information in light of building better relationships up and down our supply chain.

The need to trust our suppliers is going to increase as our business models adapt to faster change. The trust element appears to be an important part of all business relationships that need to be sustained for high performance. Clearly, contracts and the procurement process are not exempt.

Trust is a vital ingredient in high-performance contracts.

The legal framework in which we operate provides a significant, but arguably unpredictable, amount of protection to both buyers and sellers. Closer relationships between parties to a contract seem to be related to greater success in the contracting business. Alignment of parties to a contract is key to common understanding. This leads to better relationships and then to higher levels of communication. All of this helps keep the contract out of trouble. But if we start with the wrong contract awarded to the wrong contractor because the contracting process we use leads to the wrong decision, we are doomed to failure at some level regardless of what we do to build relationships. The right contractor (and for the seller, the right buyer) is one with whom we can enter into an agreement with a clear and unambiguous set of expectations that are fair and reasonable to both sides. This means we need to understand two things and do a third. The things we must understand are the legal framework in which we will operate and the nature of what we are trying to achieve. Then the thing we must do is select the best contracting strategy for the task at hand.

It is important to put the right things into any contract. Part of this is the right mix of expectations for both sides to the Agreement.

Today's business environment requires speed, focus, and customer satisfaction. This means our contracts need to be more flexible. They need to be based on established relationships whenever possible, and we need to develop exceptional communication skills.

Table 12-1 shows the steps that we should consider at each of the main stages in developing and delivering on a good contracting strategy.

Phase	Steps to Consider	Rationale
Set strategy	Use a consistent approach to contracting, especially if there is more than one contractor on the project.	It pays to have the same principles to apply to each company that is engaged on a project.
Prepare contract documents	Align contract strategy with corporate systems and procedures.	What the contract requires and what is needed internally for administrative purposes should be aligned.
Select contractor	Make sure that the selection process is fair, especially if selection is based on a bid process.	It costs money to bid on a contract. The process needs to be equitable in order to justify contractors' time and money. (In Canada, this is a legal requirement.)
Negotiate and finalize contract	Clarity and a real common understanding of the contract are required to avoid future problems.	Eliminate latent disputes. Obtain alignment using the Priority Triangle and other tools. Only if the parties to a contract have a common understanding of the objectives and priorities from the outset will the probability of success be high.
Contract administration	Avoid doing anything that bypasses contract terms and conditions.	This may change the contract by implication. For example, do not make direct payment to contractors' or suppliers' subcontractors or subsuppliers.

Phase	Steps to Consider	Rationale
Manage changes	Treat a change order as a contract within a contract. It needs to be properly documented at some point.	A successful change order is one that covers the cost to the contractor and does not cause the buyer problems. It may need to be documented after the work is done to avoid delays in, or effects on, the work of the contractor.
Resolve differences	Consider writing into the contract that litigation under the terms of the contract is not permitted, and include alternative dispute resolution methods in contract.	By eliminating the opportunity to litigate under the contract, one can still litigate in tort.
Close contract	Agree on the definition of success at the outset. Build a SMART Charter and start the partnering process as early as possible.	Do not consider a contract to be complete until all the parties to the contract are content with the outcome. It is necessary to have sustainable relationships with suppliers and clients.

Table 12-1. SMART (strategically managed, aligned, regenerative, transitional) contracting steps within the law.

12.2 Thou Shalt not Mix up the Wrong Work Packages

> *The scope and form of each contract need to be thought out carefully, as do the ramifications of a particular strategy considered in developing the way in which the project is to be procured.*

A contract starts somewhere. The starting point is normally an identified need or desire by an individual or organization to acquire something. This is such an obvious statement: Why bother even making it? The reason is simple. The success of the contract lies in how well and effectively we furnish the buyer's need. So we – as a buyer or a seller – need to understand exactly what is being bought.

We buy only three types of things: commodities, products and services. A commodity is something that differs only in price, availability or delivery times when bought from one vendor or another. Specifications, legislation, industry standards or some other third party standard may dictate the quality. If we are buying a commodity, we will be looking at the price, delivery time, and reliability of the vendor. Which of these governs should be in line with the priorities for the project.

If we are buying a product, then we are adding the expectation of a specific quality and post-purchase assurance that this implies. A product would typically be differentiated from a commodity by its branding and warranties supporting the presale promises. There is probably a difference in two physically or technically identical products if one has a well-known manufacturer and the other has an unknown manufacturer. With products, we tie the reputation of the producer to the item we are buying. There are different expectations. Typically, we are willing to pay for the real or implied warranty that a brand name implies. Consider a Donald Duck notebook computer (apologies to Walt Disney) compared with an identical machine made by IBM® or Toshiba™. Or a Yukon-Shoveit bulldozer compared with one made by Caterpillar®. Or perhaps a BMW™ car compared with one made by Feudal Motors. In many cases, we will pay more for a more reliable (or at least, known) brand, manufacturer or specialist product.

If we are buying a service, then we add one more component. For a service, we need to add the quality of the relationship. For virtually any service, whether in our daily lives (doctor, dentist, mechanic, plumber) or at work (accountant, lawyer, engineer or architect), we need a relationship with the person or people with whom we are dealing. Generally, the better that relationship, the better we feel about the quality of the service we are getting.

How we select our vendor should be based on whether we are buying a commodity, a product or a service.

As a buyer, we always have the choice of what we will buy. Sometimes, we can even choose how we buy an item. For example, we can choose to buy brain surgery as a commodity. In this situation, we simply select the least expensive surgeon who makes a claim that he or she can deliver what we want. If we elect to treat this as a qualified commodity, we may screen those brain surgeons who have a better-than-60% success rate and then select the least expensive. Alternatively, we can choose to treat this as buying a product. In this case, we may go by reputation of the available surgeons, selecting the one who is most respected for the type of operation we are looking for. In this case, money and, possibly, time will be less important. If we choose to select a surgeon as a service, we will perhaps shortlist a few based on reputation and then likely select the one that we feel most comfortable with – maybe after an interview.

In this example, there is a loose correlation between the choice of what we are buying and the priorities that govern the choices. If money is the first consideration, buying an item as a commodity makes some sense. If we are concerned about the product, then time and performance are more important. If we are buying a service, we are probably looking for performance and an intangible level of comfort first. Time and cost will be secondary considerations.

How do we buy engineering services? Or other services? How about when we buy a product? Just how many of these do we buy as a commodity and expect anything better than just what we have bought?

If we buy anything as a commodity, we cannot necessarily expect service or reliability as part of the deal.

Sometimes we will select a contract type because it is what we normally use. However, the effective type of contract will be dictated by how well we can define what we are buying. Here is an example:

We need an updated procurement management system. There are a few things we know we need, and we have written a general specification for it. The specification is in terms of what the system will do to facilitate the procurement function. It also includes a requirement that it is seamlessly integrated with our electronic data interchange system, the accounting system for accruals, general ledger, accounts payable, and other functions. Finally, we want it tied into the corporate inventory management systems as well. We decide that we want this bid as a stipulated price contract with four invited bidders. We award the contract to the lowest bidder. What have we done? First, we have procured the system as a commodity. This means that anything that is likely to require any additional effort, true integration with the other systems or a redefinition of the scope for clarity purposes will engender a change to the contract. With a performance specification – and the need to tie into several existing systems, which will likely have been modified or tailored by someone somewhere – we will have many changes. Effectively, we have entered into a cost plus contract or a unit rate contract if we have a set of rates for the various specialists and other personnel that the contractor will be using. In other words, the type of contract is not aligned with what we are contracting for and the extent to which it can be defined at the time of tender. It is really important that the payment mechanism is in line with the extent to which we can, and have, defined what we are buying.

If the work or product is defined precisely, then a stipulated price is appropriate. If the contract is likely to span more than 3 months, then progress payments are likely to be necessary. If the work involves a service, the progress payments may be appropriate even for shorter durations than 3 months.

If the nature of the work is well understood, but the amount to be done is unclear, then a unit rate contract may be appropriate. Undefined work, or incompletely defined work, should be awarded on a cost plus basis.

Again, the buyer has a choice. A buyer can do what it wants in selecting the type of contract it uses. However, be aware that the less aligned the contract type is with how well defined the end result is, the harder it will be to manage after award.

The way we pay a vendor should depend on how well the product that it is delivering has been defined at the time of contract.

That's the project over there.
Here is where we process the contractor's payments.

So far we have seen that, as a buyer, we can decide whether to buy a commodity, a product or a service. We have also seen that the type of contract – and, hence, the payment method – should be aligned with the degree of definition of the end result. Now we can decide on how we bundle what we want to acquire.

If we have the capabilities to integrate all of the components of the end product ourselves, we may simply buy the pieces. However, as companies focus more and more on core business, these skills are increasingly being outsourced. So we may need to buy the integration skills from a vendor. It is important that we package the work to suit the skills of available vendors, whenever possible. If, for example, we need to finance a project, acquire some engineering, and obtain four distinct pieces of equipment and the specialist labor to install them, how do we package this and contract it out? If we believe that we can get a better financing deal if we use our own resources, we may arrange for the money in-house. If not, then we would include this in a deal that may take the form of a leaseback, or we may leave the financing options open to the vendor. If we have the ability to manage the engineering and then the procurement and installation of the equipment, we may package each of these items separately. If we are not in the business of engineering, and it makes sense to outsource everything else, we may go to vendors of the equipment and ask them for proposals. The scope could include providing all of the equipment, providing the engineering necessary to deliver what is required and then installing the equipment and financing the whole package. This could be a turnkey contract or may be a build, own, and transfer (or BOT) contract. The key, however, lies in finding companies that can do this better than we can.

 How we bundle the goods or services to be provided under a contract should be based on how well we can define what we want; on whether we are buying a commodity, a product or a service; and then on the capabilities of available vendors.

We have started with the buyer. Without a buyer, there is no contract. And, in the end, the whole idea is to have a happy buyer and a profitable vendor. If we have these two goals in mind at all times, we are likely to put a more intelligent contracting strategy together. For the buyer to end up happy, it needs to know both what it wants and what the capabilities and potential of the vendor are. There is no point in asking a vendor to do something that it cannot hope to achieve. The buyer will be disappointed. If there is no vendor that is qualified, then the buyer needs to decide if it is willing to pay for the training of that vendor so that it can provide the needed service.

 If we start with the needs of the buyer, then the buyer should understand two things: what they really need and what the potential vendor can provide.

A good buyer will not ask a vendor to provide something that it is not truly competent to provide.

From time to time, contracts are written in which the vendor is asked to effectively protect the buyer against risks that it has no real control over. These risks can include many different things. Some of the more common ones include the following:

- Extraordinary weather conditions
- Strikes by unions
- Existing (but hidden) conditions at the site of the work
- The state of existing facilities, services or systems
- Consequential damages
- Future changes in regulations
- The skills of the operators of the finished facility or system.

Most vendors are not in the insurance business. As such, they do not have the actuarial skills to assess the risks they are sometimes asked to take. The result is that the premiums attached to these types of risks are somewhat arbitrary and are more likely to be based on the person making the assessment and his or her experience or opinion rather than being based on any science.

 One of the important elements that vendors are poor at providing is protection against risks that they do not control or cannot reasonably influence.

All too many clients ask for solutions to a solution. It pays a vendor to find out what the underlying problem or challenge is. That way, the opportunity to meet the client's needs increases, as does the opportunity to offer a better solution, should such a thing exist. While we are talking of better solutions, what the vendor believes (or knows for certain) is a better solution, may not be a better one in the eyes of the decision-makers in the buyer organization.

From a vendor's perspective, it pays to really understand what the customer wants. They want a solution to their problem.

The difference between a right solution and a wrong one may simply be attributed to politics within the buyer organization. As a vendor, it pays to understand these points, as there are three things a typical buyer is NOT interested in. The first of these is any problem that the vendor may have. Vendors are there to solve customer problems. They are not there to complain about how difficult life is for themselves. Any vendor that delivers an unpleasant surprise is not very popular with the customer. It pays to underpromise and overdeliver. Finally, no buyer likes the wrong solution – maybe even (or especially?) if the solution is a better one than they thought of and took time and trouble to sell in-house to the boss.

It pays for a vendor to understand what the customer does not want. Three of the things most customers do not want include the vendor's problems, embarrassing surprises and the wrong solution.

To summarize . . .

The strategy for contracting on a project and the choice of contract we use should be based on what we know about the project, our skills, and what we can acquire on the market at the time.

How to reduce anyone's productivity quickly!

Now that we have a basis for determining a contracting strategy, we need to look at how we select the vendor. This is tied in large part to what we are buying. Selection of vendors for commodities requires little skill: We simply select the vendor that can deliver in the required time frame at the lowest possible cost. For products and services, however, it takes a bit more expertise and effort to select the best vendor.

For selection of a product, we need to consider the other elements that led us to buying it rather than treating it as a commodity. For example, were we concerned about after-sales service? Or reputation? Maybe reliability? Maybe we wanted a level of assurance that what we were buying will predictably do what it is supposed to do. Whatever the reasons, we need to identify them and attach a value to them. For example, reliability may be assessed as being worth an equivalent to 100 hours of operating time per year saved. This additional productive time may have a net value to the business of $45,000 per hour, so a reasonable premium, with a payback period of 1 year, would be $45,000 × 100 hours = $4.5 million. In other words, on this basis, we can justify spending up to $4.5 million MORE than the cheapest product if the difference is the level of performance improvement that is equivalent to 100 hours of additional production per year. This is a crude calculation that needs to be tempered by probabilities, cost of capital and other factors.

If we are buying a service, we must consider what it is we need from this service. If, for example, we are looking for combined engineering and construction services, what are the factors that are important to us? How do we rank and weight them? Let us look at an example:

Consider a manufacturer of automobiles. The company wants a design and construction package to rebuild a part of a car assembly plant. The plant is being modified to produce a new model of car, and the construction involves relocating a painting booth and four large robotic welders. In addition, there is a need to fill several pits and trenches required for the old assembly line and to cut and form three new ones to accommodate equipment to be provided for the new line by another vendor. The work is to be done in a short time to coincide with the plant's annual shutdown. The vendor has decided that this is a service. The work is not well defined yet, as the details of the new model and its assembly are still being finalized. The service needs to be engaged now if the project is to be completed on time. There are two vendors that have done work of this nature before for this client. There is a third vendor that the client would like to consider as well, although the client recognizes that there is some risk associated with taking on an inexperienced supplier for this critical project.

When we analyze the issues, we find the following criteria to be important:

Item	Relative Weight	Value
1. Knowledge of the plant	100	$50,000 (A)
2. Understanding of client engineering procedures and systems	150	$85,000 (A)
3. Experience in this type of work	200	$150,000 (B)
4. Relationship with other vendors and knowledge of their products	50	$20,000 (A)
5. Relationship with client team for this project	250	$100,000 (B)
6. Construction capability	200	$1,000,000 (C)
7. Bid price	100	-
8. Proposed method of working	250	-
9. Schedule for the work	180	$200,000/day (D)

The assessment of the value is based on the following:

(A)   The cost in vendor charges to bring up this item to the level of the best vendor
(B)   Estimate based on the experience of the project team
(C)   Estimated value of the risk associated with this item
(D)   Lost profit per day of production

We have now turned the criteria for assessing the vendors into a basis for making comparisons between the bidders. Each of the bids can then be scored against these criteria on a scale of 0 to 10. For example, bidders A and B may have the same knowledge of the plant, but they lag behind contractor C, who is assessed (based on who is assigned to the team in their proposal) as having complete knowledge. Contractor C scores 10. The other two score 5 each. The score of 5 is based on an assessment of the cost of bringing the team up to the same level as contractor C, being half (5 out of 10) of the value of $20,000.

Each of the criteria is assessed in a similar way. The total score is then calculated, and the bid price is adjusted to reflect the different values associated with the items being assessed. The overall best deal can then be determined in an auditable fashion that does not rely just on the lowest initial price. Granted, part of the process is subjective. I did not say this was easy. You need experience, knowledge and applied intelligence to make a more sophisticated assessment work.

Set the contractor selection criteria in order to pick the best overall vendor for the situation, not just the cheapest supplier – they may not be the same.

Whether we are a buyer or a seller, it usually pays to screen our potential partner. As a buyer, we want a reliable, flexible and honest vendor. As a vendor, we want a client that knows what it wants and can afford to pay for it. Likely we have other criteria as well. The point is that if we find a partner that does not meet these criteria, we either look elsewhere or add whatever safety nets we need to the deal. As a buyer, this normally takes the form of added conditions in the contract and additional supervision or management. As a vendor, this takes the form of added costs of the precautions we will need to take, including additional management costs, contingencies and premiums associated with uncontrollable risks.

There are some options to consider in establishing the most appropriate process for selecting your vendor or for screening your potential customer.

Before signing a contract that then becomes a private law between us and our business partner for the duration of the agreement, we both need to be sure that we clearly understand the contract and that it MEANS

THE SAME THING to both of us. Chapter 5 explained how easy it is to build latent disputes into our contracts. To avoid problems, it pays to have a discussion about success. I have found it inordinately useful to have both the buyer and the seller picture a celebratory party at the end of the project. Then ask some questions: Why are we celebrating success? Who is doing the celebrating? What are they saying? We should consider the answers and see if there is any conflict between them. If there is conflict, we need to address it before we sign the contract, changing the contract if necessary before penning our signature.

 Always be sure that you and your business partner in a contract have a common understanding of what the contract is about.

12.3 Thou Shalt Listen to, and Understand, the Real Wishes and Needs of the Client

> *Strange thought: Look after the client. Its real needs and priorities will dictate the contracting strategy employed.*

One project manager I know quite well had a reputation that spanned the industry he worked in. Why? Because of one specific habit: He was dictatorial in how he wanted things done. He would not stand any "nonsense" from his contractors, especially his engineering contractors whom he would command to do things his way. After all, he was the client – and the client is always right. If – and it did happen from time to time – it turned out that his ideas were wrong, it was still the contractor's fault. The contractors would typically point out that the decision was the client's. In response to this, the project manager would confront his poor contractor and say, "Well, you are the expert. I hired you to do it right. Why did you not tell me at the time that this was not the best solution?"

To be effective, there is always at least one key piece of information that must be shared. Unfortunately, it is not always conveyed fully and accurately. As a result, we do not achieve proper alignment of objectives and the contract does not run as well as it should. What is that key piece of information? It is what the client is trying to achieve. This may be something bigger than, or different from, what the contract is intended to deliver. Knowing this helps the contractor understand the contract and what it needs to deliver in a business and social context, as well as in the technical one we are so used to. It may even lead to a different outcome.

 It is a myth that the customer is always right. Sometimes the customer just THINKS it is right.

In some organizations, the procurement process is separated from the end user of what is being purchased. The greater this separation, the more difficult it is for everyone to work effectively. The reason is simple: The further the end user (beneficiary) of a contract's output and the contractor are removed from each other, the more difficult communication between them is.

Typical symptoms of this remoteness include increased numbers of changes to the contract, miscommunication and unmet expectations resulting in acrimony, blame assignment, non-cooperation and high stress for all. The cure lies in shortcutting the formal channels and establishing the necessary relationships between the end user and the contractor's staff. The challenge with this approach is managing everyone's expectations so that the administration of the contract remains in alignment with what is happening. The best way to do this is to encourage the procurement team to manage this communication by initiating it and adapting the corporate systems to accommodate free flow of information with the appropriate controls in place to allow changes to be managed effectively. The earlier the communications are established, the better.

The earliest opportunity for such communication between the purchasing staff and the end user is BEFORE the contract is assembled and the strategy set.

Particularly in large organizations, the procurement process can be too far removed from where the problem exists for the contractor to understand what it is trying to do from either a business or a technical perspective.

Here are a few examples of disconnects in objectives that happen all too easily if the end beneficiary and the procurement process are separated.

One buyer wanted a parts distribution warehouse. He hired a contractor to design and build the facility. The contractor went to the root of the problem and suggested a different solution. The new solution lost him the contract, but it won him a lifetime client. The solution saved the cost of the land and the building. It eliminated the inventory that had to be carried in the new (now nonexistent) warehouse. The solution was to use an existing warehouse in a different location and charter a regular cargo flight from there to the city where the new facility was to be built. The end result was a better business solution for the real problem.

In another situation, a large North American city was about to build two projects. These projects were the culmination of over a year of study. One project was budgeted at just over $2.4 million and the other at about $5 million. When the problem that these projects were really addressing was revisited, a different solution was quickly found. The new solution was based on different technology. The combined value of the new solution was less than half the original budget and could be installed in a fraction of the time. There are many examples like this. The good ones end with the best solution being found before the wrong one has been built. And there are many examples like that, including a $400 million (give or take a bit) magnesium plant in Alberta, Canada, which was never operated.

 Sometimes it takes some digging to unearth the real problem that the customer is trying to resolve.

It really pays *all* parties to a contract to understand what the buyer organization is really trying to achieve.

So much for the wrong project. What about the wrong contractor? Have you ever ended up with the WRONG contractor as a result of a standard and accepted method of selection? Not overtly and obviously wrong, necessarily, but you may have preferred to see another contractor win the deal. Maybe your instinct was that this was a bad decision, and this instinctive sense turned out to be justified as the project progressed and problems grew, delays stretched out and costs increased. If this is the case, you are not alone. Sometimes the rules we make for selecting a contractor get in the way of selecting the right one. And as a result we end up with problems that we could see coming. The selection process should be designed to eliminate inappropriate contractors from the outset, without limiting your choices and the competitive field in which this is needed.

Consider prescreening prospective contractors. (Most of us do this anyhow.) Also think about an entry-level open bidding option. This allows any contractor to bid on small or noncritical projects. If they are successful, then they can earn the right to move up to the next size or level of project. We may want to have several levels of contracts that require increasing experience with our organization, growth in project size complexity and so on.

Make sure that the contractor selection process does not get in the way of solving the real problem.

Whether we are the buyer or the seller, it is important to manage the expectations of the other party. The definition of a successful project is that all the stakeholders are happy at the end. So we need to know – and care about – the realistic expectations of the other party. For example, if we are a contractor and we expect to be paid in 14 days but the contract says 30, we must discuss this with the buyer and make sure we can get what we are expecting. If this is not possible, the expectations need to be managed by both the contractor and the buyer. Similarly, if we are the buyer and we want to accelerate the project but do not expect to pay for this, we must discuss this with the vendor and make sure it can be done before our expectations end up not being met. Remember, the other party cannot deliver on an expectation that has not been articulated.

We must manage our partner's expectations. This starts before contract award and ends after the project is complete.

SMART Management™ provides a number of tools to help articulate and manage expectations. These include the Priority Triangle, Three Key Questions and project classification scales. These and other tools are described in more detail in the book "Don't Park Your Brain Outside" (Hartman, 2000). Following is a brief description of each of the three tools just named.

Priority Triangle: This simple tool is an inverted triangle. It is shaded such that only six spots are left white.

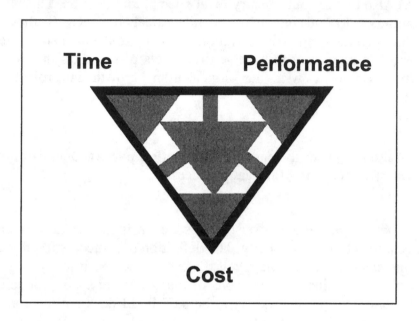

Figure 12-1. Priority Triangle

The idea behind the triangle is to agree on where to place an X. It must go into one of the nonshaded areas. The closer it is to a corner, the more important the element assigned to that corner is. So, an X in the upper left corner would identify time as most important, followed by performance and then cost. The management of the project and the contract would then reflect these priorities. Now that we know the buyer's priorities, what are the expectations around completion and success?

The answers to the **Three Key Questions** tell us three things. The first question is **who** decides on the answers to the next two questions. This is normally the person or persons who will determine whether the contract was successful or not. It pays to understand who these people are. The next question provides the answer to what the **metrics for success** of the contract are. These may be the start of production in a plant, improved customer service for delivery of a computer system or a successful launch of a new product for an advertising contract. The answer to the third question tells us what the **last deliverable** is, that brings the contract to a close and provides agreement on who is responsible for it, as well as what is needed so that the last deliverable can be produced.

Make sure that everyone knows what success and completion look like for the contract – and that they all "know" the same thing.

The third tool looks at buyer and seller expectations from yet another point of view. The **classification of the project** addresses five dimensions to help both the buyer and the vendor better understand the project at hand from the perspective of the other party. These dimensions are:

- Size: This is the effect that the contract has on each party's business. What may be a trivial project or contract for one party may be a massive undertaking for the other.
- Complexity: This is a measure of how many different skills, products, companies or other stakeholders are involved that need to be coordinated. Again, this may be different for the parties involved.
- Uncertainty: This is a measure of how well the end state of the contract can be defined at the outset. The greater the uncertainty, the more flexible the contract needs to be to accommodate changes or evolution, as the case may be.
- Constraints: This is a measure of the flexibility that each party has to accommodate change. Perhaps there is an inflexible end date. Or maybe the specifications must be met exactly, with zero tolerance. The budget may be severely constrained. Possibly all three of these constraints apply. Whatever the case, discussion will allow the parties to determine how to deal with a problem or crisis.
- Ugliness: This is the final measure. It is an assessment of how attractive this project is to the participants. An ugly project will likely have high stress and staff turnover. It is often better to understand this from the outset, so it can be managed.

The more these points can be discussed, and the earlier the discussion can be take place – ideally before signing the contract – the more likely the parties are to be aligned and, therefore, successful in delivery of the end result.

We must take the time to get aligned with our contracting partner. We should use the priority triangle, three key questions, project classification, and other tools as needed.

What I have just suggested does not fit many cultures where the buyer plays the role of master and the vendor is the servant. Today's world is one in which collaboration rather than confrontation is needed to gain a competitive edge. The more we come to rely on our supply chain, the more important the relationship between buyer and seller becomes.

It is not enough just to gengage a contractor. Buyer and seller need to work together for success. Share and agree on how you will manage your joint project and who can make decisions for whom about what.

12.4 Thou Shalt not Blindly Pick the Contractor who was Cheapest Because it made the Biggest Mistake

> *Contractor selection is not just a matter of selecting the lowest initial price. Longer-term thinking usually pays off.*

Section 12.4 reinforces some points about contractor selection. The first point is that we need to tailor our process for contractor selection to suit the situation. If we normally competitively bid our work, and we need a specific contractor quickly for a highly specialized and urgent task, what do we do? Bid the work and disregard the huge cost of the delay that results? Unlikely. We already do some tailoring in almost every organization, although this tends to be driven by crisis. We want to be more proactive than that.

In many organizations, we have an almost sacred contractor selection process that is nonspecific to the problem we are trying to solve or to the situation we are in. This is a problem only if the process assumptions do not exactly meet the specific criteria for our project. Unfortunately, this is most of the time!

We can justify almost any deviation from company policy in a crisis. But managing changes to established procedures becomes harder when there is no panic driving the decisions. Unfortunately, it would be inappropriate to suggest creating a crisis in order to effect change – although this has been done in the past. Instead, consider increasing the options for contractor selection, based on the characteristics of the project. Specifically, look at the five-scale classification system discussed in Section 12.3, and see how each one would affect the selection of a vendor.

We should be able to adapt the vendor selection process to suit the specific needs of our project.

This is worth repeating:

Selection of the right vendor should only be based on the lowest bid if the only consideration is money. (You are buying a commodity.)

And so is this:

When you are buying a product, consider – and try to put a value on – the vendor's reputation and the real and implied warranties.

And, finally, this:

When you are buying a service, your relationship and ability to work with the service provider are as important as its ability to provide the required expertise. Both of these are more important than whether the rates charged are the cheapest on the market.

There should be two distinct steps in a bid or negotiation for a contract. The first is to screen vendors. The vendor should be able to do the work, have the business skills to survive the contract and have the management skills to deliver what we need in a timely and cost-effective way. If the contractor passes this screen, then make sure that he or she understands what is required and can positively commit to delivering what we want. These are not the same thing. The first step is an assessment by the buyer of the vendor. The second step is an assessment by the vendor of the buyer and his or her expectations.

Prequalifying vendors and determining whether they have truly understood what the contract is about are not the same thing.

Know what is important for success of the project and set the selection criteria to suit the occasion.

As buyers, we are often a bit nervous about showing our hand to our vendors. There are things that we consider to be proprietary or that may lead to higher prices if we disclosed them too soon. We need to ask ourselves whether we are buying future problems if we do not fully disclose our intentions. Similarly, if we have established selection criteria, we should tell the vendor what they are and how they will be applied. We can give examples when appropriate. Only through open communication from the outset can we hope to exchange the information we all need to be most effective. No vendor can hope to give us what we have not asked for, unless it is by chance.

We must make sure that all bidders understand exactly what we are looking for, so that they can present their skills and resources accordingly.

What I have just suggested is that we trust our suppliers of goods and services. We should trust them, but put the checks and balances in place that are prudent for the situation and the vendor. The need for checks and balances should decrease as our relationship with the vendor matures. The relationship will mature and become closer as:

- We get to know each other better
- The relationship survives more contracts
- The vendor becomes more dependent on us for survival – or *vice versa*
- There are growing advantages to both sides to sustain the business relationship.

 Trust, but validate!

12.5 Thou Shalt not be Ambiguous and Vague

> *Ensure that the Contract is clear and unambiguous whenever possible. Add a step in the process to make sure avoidable disputes are eliminated.*

SMART Contracting is all about getting what we need from a third party, and getting it as effectively as possible. There is more to this statement than meets the eye. The very first question to ask is: Do we really know what we want? I know this is a cheeky thing to ask, but we do not always really know what we want because it has been defined in terms of a solution. I may tell you I want a warehouse, but what I REALLY want is faster delivery to my customers, or guaranteed turnaround times for orders, or lower inventory costs or greater storage capacity or perhaps the ability to support more products with spare parts.

There may be better, more cost-effective solutions out there. Even if we do want a warehouse, should we build or rent? If we build, is the design based on our standards and experience or could we learn from the competition by hiring the designers or experts who have been exposed to their operations?

It is usually better to describe the problem to be solved rather than trying to describe the solution, unless we are really sure that our solution is the best one.

Performance specifications tend to move closer to the problem we are trying to solve rather than any preconceived solution we may have in mind. This is a good thing. Even if we know exactly what we want, we should consider building in as much flexibility for the vendor as possible. This allows the vendor two important things: First, it permits a degree of creativity in bringing the cost and time required down through suggestions of alternative materials, methods or other options. Next, the process also allows a degree of ownership and commitment that just complying with the requirements does not quite deliver.

We underutilize performance specifications.

One of today's challenges lies in knowing where the real expertise lies. Who is best placed to understand how a pump is used, what can go wrong, what the options are, differences in reliability or performance under different conditions and how to build or maintain that pump? The engineer who specified it or the manufacturer who built it, serviced others, helped other users of the same product under different conditions and addressed warranty issues? Put this way, it would seem that the pump manufacturer should know more.

The role of the engineer used to be to design and specify the components that are required to serve the needs of the client. This has gradually changed over the years. Today the design engineer fills the role of the integrator. Many client organizations have outsourced most of their engineering function. Some still handle maintenance and minor upgrades in-house, whereas others have contracted out even these functions. Operators know what works for them, but may not be aware of the latest technological advances. Yet all of the operators, maintenance specialists, the people who have made endless changes to what was originally built, suppliers and contractors have something to contribute to a good solution for a given problem. The best engineers today harness all of this expertise, add their own knowledge and experience and distil the result to address the client's problem or need. It is increasingly common to leave the details to the equipment supplier, installer or constructor. We design the detail by shop drawing. The same principle holds true of information systems, lawsuits, advertising campaigns and just about any other service we require in business today. Perhaps the exception is a very specialized expertise needed for a specific task. But then we typically go directly to the vendor we need.

In engineering-based projects, the role of the engineer has shifted over the past few years from being a solution provider to being a solution integrator. The difference lies in where the detailed and highly specialized product expertise is: It is with the vendors.

There are times when we need to specify exactly what we require. There may be issues of compatibility with existing equipment. We may have a policy that limits the number of vendors and, hence, the amount of spare parts we need to carry and people we need to train. There may be safety or regulatory requirements that have to be met. Whatever the reason, we end up with a very specific and prescriptive specification. But we rarely, if ever, explain why this is the case. If we did, we may be surprised at the opportunities to improve the end result that might emerge. It could be because someone further down the supply chain knows of a better way to solve the problem.

When we do provide specifications, they should be clear and unambiguous and should explain why a particular approach has been taken.

If we now start inviting input to the design by others, we still need assurance that the end result is technically sound and is not a camel designed by a committee trying to improve on the horse. This is managed in a number of ways. One effective approach is to process all suggestions through one authority. Whoever is the designated authority is likely to be the designer of record. As engineering and architecture are regulated professions, this role carries significant responsibility. The designer of record is normally ultimately responsible in law for the overall integrity of the design. As such, the designer must have veto power. If the fee structure is not designed to allow the necessary work to be done to ensure the robustness of the end product, expect the designer to err on the side of what he or she knows and therefore feels safe with. In the end, conservative or overdesigned solutions can cost significantly more in both capital and operating costs than the additional engineering or design fee that was saved.

Managing a program for design, procurement and installation or construction requires careful alignment of the various contracts so that the requirements of one do not interfere with the efficacy of another.

 Integrate contracts and roles carefully.

We have just dealt with one aspect of risk in assembling contracts. There is a tendency, especially in North America, to divest oneself of risk. This starts with the end customer and tends to be passed down the supply chain until it comes to rest with the last contractor or subcontractor. If we pass on inappropriate risk, we should do so with some careful thought as to where that risk may end up and what is likely to happen in the event that the risk event occurs.

Before the risk occurs, the chances are that the vendor asked to take it on will do two things: First, the vendor will add a premium in some form, as discussed in Chapter 6. Then he or she will try to pass the risk on to the suppliers whenever possible. Then the suppliers will add their premium . . .

And then the risk event occurs, and everyone runs for cover. Interpretation of the contract and assignment of blame become the flavors of the month. We still need to go through the pain of resolving a problem that will likely end up as a dispute.

 Do not ask vendors to take inappropriate risks – this adds cost but rarely saves the buyer any grief.

If we know we regularly have differences with our client or with our suppliers, we should do what we can to minimize the incidence of such events. We have just looked at problems generated by inappropriate assignment of risk. Now we can use the risk approach to test whether we have a difference in interpretation of a particular clause. It pays to take the time to root out any latent disputes.

 We can avoid latent disputes by taking the time to obtain clarity and uniformity in our understanding of the Contract with the other party to that Contract.

Latent disputes are relatively easy to identify if we use the approach outlined in Chapter 5. Latent resentment or bad feelings are harder to define and address at the outset. Most experienced buyers and sellers of goods and services will have had the experience of a contract in which there was a sense of resentment. It takes the form of slow and surly responses, anger, attacks of the other party's credibility, performance and staff, arguments and a general attitude shift that is counterproductive. Can this be stopped? The underlying cause of such behavior can probably be linked to a sense that the contract terms are unfair to one party in most of these situations. That the party in question entered freely into the agreement is neither here nor there. The problems arise when the real or apparent unfairness starts to damage profitability, performance and possibly careers or prospects for the people trying to deliver their part under the contract. A truly fair contract goes a long way to reduce the risks. The question that remains is whether or not the "fairness" is at the expense of commercial astuteness. Again there is a need for balance, as the problems associated with unfair contracts may be greater than the advantages we may gain by enforcing them.

Fairness in a contract goes a long way to reduce conflict and, potentially, overall project costs.

As part of the process of achieving a fair contract, getting commitment to the terms and conditions by both parties is probably the most significant step. This commitment has value only if all concerned are truly committed – and this is based on understanding the same thing. In other words, if we interpret the Contract one way and the other party interprets it differently, but neither of us knows that we have committed to a different interpretation, we have a potential problem. The problem will manifest itself when something triggers a need to interpret that Contract. Then our different interpretations will be well set and relatively intractable. This will make resolution of our difference hard. So we should work at resolving these problems ahead of time. It pays to take every opportunity to clarify the meaning or intent of a contract before it is signed. If the contract is being bid, create opportunities before and after the bids are submitted to clarify and address issues. As the bidders look at the documents, they will have questions. These will help identify problems in interpretation, and we should fix them before we close the deal.

 Take the time to clarify the terms and intent of the Contract both before and after the bids have been submitted.

Once we have closed the deal, we need to remember that all we have done is set the ground rules – the private law – that will govern our behaviors and negotiations after that event. Signing of the Contract is simply the beginning of phase 2 of our negotiations. Phase 1 sets the basic rules for payment, work, delivery of goods and services, quality and safety standards and other special conditions.

Remember that when you sign the Contract, it is the beginning of the next phase of negotiation, not just the end of the last phase.

Even when we have signed the Contract, and the deal is well articulated, we know that things will change. There should be appropriate allowances for this. The vendor should have allowed in its rates or its price for the things it is liable for under the Contract. This will typically include, for example, its own performance and the quality of its goods or services, any actual or implied warranties and the wherewithal to make up time lost as a result of its own errors or omissions.

Similarly, the buyer should have a budget larger than the price of the Contract. This budget will include for the additional costs and other problems associated with its need for change, its own performance or failures under the contract and risks it has retained. It is a foolish buyer that budgets only for the initial face value of the Contract.

> **The difference between the successful vendor's price and the budget of the buyer should be the contingency that the buyer carries to cover its risks under the Contract.**

12.6 Thou Shalt Share out Risks Equitably and with Intelligence

> *Risk apportionment needs to be done intelligently. Section 12.6 presents one effective way to apply the Calgary Method.*

The end of Section 12.6 explained how we need to carry a contingency over and above the price of the contract (for the buyer) and the cost of the contract (for the seller). The cost of the contract is the estimated cost of doing the work. A vendor will add to this any additional money needed for overhead, profit and risk. Overhead is dictated by how the vendor does its business. Market conditions and other factors dictate profit. The only item we can do something about on a contract-by-contract basis is the risk and the premium associated with it.

Similarly, the wise owner will add a contingency to its budget for a contract. This contingency will be earmarked to cover the anticipated cost of the risks associated with the contract. We can influence this sum too. But to affect these risk-driven contingencies, we need to understand and assess the risks we are dealing with.

> **It invariably pays to have risk assessed and understood before developing the contracting strategy.**

It makes a lot of sense to assign risk to the party that is best able to manage it. The premium should theoretically be lower. This is explained further, below, as theory and practice are not necessarily the same. If we are to assign risk to the appropriate party, we need to consider this early in the contracting process, as it will affect the strategy we adopt.

As it turns out, our profit calculates as 100%
minus the percent completion on any day.

 **The contracting strategy should reflect the risk management
program for the project.**

The first step in good management of risks in procuring goods and
services is to pick the right form of contract. Stipulated price contracts
imply a high degree of certainty regarding the end product. Cost plus
contracts imply low certainty. The risk for definition of the product lies
with the buyer initially. After all, we cannot buy something if we do not
know what it is. At least, we cannot agree on a price until we can define
exactly what we want from the vendor. So, if we need to start work before
we can finalize the details, we will need to take the project definition risk
back as a buyer. If we can define the product, we can pass the delivery risk
on to the contractor.

 **Pick the most appropriate form of contract to meet the needs
of the project.**

Once we have defined the basic form of the contract, we can start to look at all of the risks and the risk-assigning clauses in the contract itself. The Calgary Method has successfully been used to assign risk based on the best commercial deal in work that is bid or negotiated. Here is how it works in a Stipulated Price contract bid situation:

1. Identify all the clauses that do, or appear to, assign risk to the contractor. (We should include those clauses that the contractor identifies as assigning risk, even if we do not see them as such.)

2. Ask the contractor(s) providing a price for the work to price the risk associated with each clause separately. In other words, ask them to say how much they would reduce the price if the risk-assigning clause were to be amended or deleted, so that the buyer were to take all of that risk.

3. Assess how much we, as the buyer, would allocate to that clause if we were to take the risk.

4. If our premium for taking the risk is lower than the contractor's, then it will pay for us to take the risk. If it is higher than the contractor's, we should pass the risk on.

5. When necessary, after clarification of what the risk is, rewrite the clause so that there is no doubt who carries the risk for the lower premium.

6. In a competitive bid situation, repeat these steps for all of the bidders, and then select the bidder using the adjusted price (see Table 12-2) and any other criteria involved in picking the winner.

Distribute risk taking such that the premium associated with the risk is as low as possible.

Assign risk only where it can unequivocally be assigned.

#	Clause	Premium Contractor (A)	Assigned Buyer (B)	Lower Premium (C)	Party to Whom Risk is Assigned
1	Soil conditions	$200,000	$270,000	$200,000	Contractor
2	Delays to work	$250,000	$100,000	$100,000	Buyer
3	Consequential damages	$400,000	$5,000	$5,000	Buyer
4	Compliance to codes	$500,000	$10,000	$10,000	Buyer
5	Third-party liability	$120,000	$5,000	$5,000	Buyer
6	Weather	$50,000	$250,000	$50,000	Contractor
7	Labor disputes or disruption	$10,000	$200,000	$10,000	Contractor
8	Delays in materials deliveries	$25,000	$25,000	$25,000	Contractor
	Total risk premiums			$405,000	Sum of column C = D
	Contractor's base bid (no risk clauses included)			$13,200,000	Base bid = E
	Adjusted shared risk bid price			**$13,605,000**	D + E = F
	Contractor bid with all of the contractor's risk premiums			$14,755,000	E + sum of column A = G
	Saving if owner premium not exceeded by actual costs			**$1,150,000**	G - F

Table 12-2. Example of the Calgary Method for contract risk assignment.

What the Calgary Method allows us to do is justify and price the risks and how they are managed. It delivers the best combination of risks and premiums, and it does so while clarifying who really has a risk. Later arguments are eliminated. The risks are assigned based on a fair assessment of what they are worth to both parties. The best commercial deal is arrived at with the most appropriate party taking the risk. The premiums are visible, and the cost is minimized to everyone's benefit.

 Consider using the Calgary Method for risk assignment on the next project.

12.7 Thou Shalt Trust thy Contracting Partner, but not do so Unreasonably

> *Use knowledge of trust types and their effect on your contracts to manage them better.*

No. I am not advocating blind trust. I am, however, trying to encourage an intelligent use of trust to help manage contracts and business relationships more effectively.

 We said it before, here it is again: Trust, but verify.

Trust starts to take shape between parties to a contract as soon as there is some communication between them. One of the earliest communications in writing is the contract or bid document or a request for proposal or quotation. The buyer or someone representing the buyer typically prepares this. Whoever prepares it tends to write a document that favors the authoring party. As a result, it tends to contain language that may be offensive to the other party. Yes, we live in a cruel commercial world, and feelings don't count. But they do. When we perceive risk, we adjust our behaviors. One of those natural defensive behaviors is to protect our own interests. Doing so costs money. Vendors only get paid by their clients, so this extra money finds its way into the client's bill in the end. This could be an avoidable cost. We know this from the studies already done and described in Chapter 6.

 Watch the language of the Contract – it may convey an attitude.

When we enter into a contract as a buyer or a seller, we enter into a regulated relationship. The contract is the regulating law that we have created for ourselves. If our contract is built on blind trust, we will likely end up in trouble. We have always known this, so over the years we have developed increasingly "watertight" documents that basically have become zero-trust based.

Both our suppliers and our clients are partners in our business. We need to treat them as such. True partnership allows either party to make binding decisions for the other. This is an extreme that most of us are not prepared to go to. However, the more we can be trusted to look after our partner's interests, and *vice versa,* the less we need to rely on the "watertight" contract. Such a contract generally passes as much risk as possible to the party that did not write the contract. The only governing factor is commercial reality. Unfortunately, we have seen that awareness of this reality in the form of premiums is low and when it exists, the implications are poorly understood by many buyers. So we persist in cumbersome, low-trust relationships. We should reconsider our approach if this is where we are in our practices today.

Treat both the customer and suppliers as partners, and expect them to reciprocate.

So, what does a high-trust relationship between a buyer and a seller look like? It will not be a new relationship, as trust takes time to fully develop. It will be an evolving relationship, based on repeat business, a degree of stability on the staffing by buyer and seller and the relationships and effective communication that this engenders. It allows for open discussion and sharing of both problems and solutions. The parties to the deal will consider any problem to be one that is owned by all members of the team. There will be <u>one</u> team, not two or more. The interests of the parties will be aligned when possible and will be respected and addressed when they are different. This removes many barriers to effective and high-performance work. Without the barriers, we have opportunities to reach new highs in performance levels.

Trust is a catalyst for higher performance in contractual relationships.

To build the trust relationship that allows for high performance, we need to build the type of trust that makes sense for the situation. Chapter 7 discusses three primary types of trust. Section 12.7 looks at each in turn again in the order that they most commonly are addressed or encountered.

Red trust is intuitive trust. This takes a few seconds – or forever – to form. Intuitive trust is our rapid assessment of, or an emotional response to, a first exposure to a person or situation. It is our first impression, and this impression lasts a very long time for most people. If it is somewhere between good and neutral, we will be open to some level of trust later. If it is between neutral and bad, we are less likely to be open to a trust-based relationship.

If we are open to some level of trust, yellow trust may follow. This is integrity trust. A good first impression opens us to some level of confidence that the person or the organization we have just been exposed to will look after our interests. We may test this through reference checks of their reputation with other clients or suppliers. Once we have established a level at which we feel we can trust this potential partner, we will place safeguards in our contract to cover the rest. Over time, these safeguards should be eliminated, if the level of yellow trust grows. Yellow trust takes weeks or months and possibly years to build. It needs to be continuously refreshed as people in one party's organization change. It needs to be maintained through constant communication and reassurance.

Blue trust is about competence. This is the most stable of all and probably the one we worry about the least. We can screen for competence. We can validate it through references, and we can provide for failure through warranty and other requirements. Blue trust takes years to earn and, once earned, can survive reality for some time. The reputation of a person's or organization's competence lags reality when it grows or declines, as the impression of capability, quality and performance is based on past performance, not necessarily what is happening now.

We need to know what is appropriate for each situation. Sustainable relationships need red trust. Without it we will forever be testing and doubting. This is hard on any relationship. For low administration and a minimalist approach to managing our contracts, we need yellow trust. To avoid rework, additional inspection or testing, quality assurance and other costs and to maintain confidence in schedules, we need blue trust. The three primary colors add up to white trust – the color of weddings. If the three types of trust are all missing, we get black trust – the color of funerals.

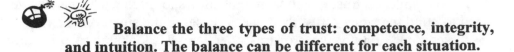

Balance the three types of trust: competence, integrity, and intuition. The balance can be different for each situation.

We can learn to assess the right balance of the three types of trust. We also need to balance the ways in which we replace or otherwise supplement high trust. Typically we use carrots and sticks to do this. Too many carrots and expectations get out of line with reality. Too many sticks, and we kill the relationship by punishing it to death. So, even when we are substituting contractual terms and conditions for trust, we need to consider the implications of the mechanisms we adopt. Penalty-based or incentive-based options exist for most of the behaviors we normally try to influence. An appropriate balance for the situation should be considered. Typically, we should move more toward incentives for positive performance when we may wish to build more sustainable relationships. In adverse and confrontational conditions, and when sustainability of the relationship is either not likely or not appropriate, more emphasis on the penalty-based approach may make sense.

There is also a balance needed between the carrot of future business and enhanced reputation and the stick built into the contract. This will vary depending on our relationship with each vendor in our supply chain.

12.8 Thou Shalt not Mess Unduly with the Contract after it is Agreed

> *Ways of avoiding unwanted contract distortions through actions are required.*

Just as in life, actions speak louder than words in our contracting relationships. From both human and legal perspectives, this holds true. If, for example, we respond to a verbal instruction and then get paid for the additional work, we will have strengthened our relationship by building trust. We are more likely to take a risk to help our partner in the future. If we accept deferred payments rather than those set out in the contract, we may well have changed the terms of the contract through our acceptance of a different set of operating rules in practice.

In contracting, actions can – and usually do – speak louder than words.

Many organizations have procedures and policies surrounding such things as making payments to contractors, selecting vendors, establishing safety standards, using company resources and so on. In working with suppliers, we will do most, if not all, of these things that are governed by policies or procedures at some point in time. If the company pays accounts monthly and 60 days after receipt and we sign a contract that requires payment every 2 weeks and within 30 days of approval, then we will run into problems meeting the contractual obligations. One government agency I have worked with has so many different levels of approval and cross checks between four ministries that the approval process alone takes longer than the time allowed in its normal contracts to make payments to contractors.

When we make the rules that are built into our contracts, we need to be sure we can follow them. If we cannot, then we should change either the rules in the contract or the way we operate. Usually it is easier to change the contract. If there is deferred payment, then we need to let the contractor know, as the contractor will have to cover the additional financing charges in its price for the project.

If we plan on managing a contract in one way but the words in the contract document are different from that way, we will create problems for ourselves. The contract should reflect how we plan to manage it.

I just used alignment of payment expectations as written in the contract with what is achievable as an example. It is just one case in which we need to maintain alignment between the buyer and the vendor. There are many other situations that will arise during the course of the contract. A well-run, regular meeting between the participants can help identify and address issues as they arise. Chapter 8 discusses the formula and agenda for such meetings. These meetings do not have to be face-to-face, although it is good practice to have at least the first one this way. A face-to-face meeting ranges from very useful to essential for establishing a real connection with the people we will be dealing with. On very large or complex projects, regular meetings are essential, as they are often the only time key people can meet, connect, and communicate. On smaller or less complex projects, a regular phone call once a week may be all that is required. However, it is important that the discipline of a regular agenda and discussion of the status of the work be had. Even for smaller contracts, the meeting minutes should be formally recorded and distributed. They are a useful, contemporary, and agreed record of progress, as well as a written document of issues and solutions, agreements, and decisions that may affect the project later.

Make effective and regular progress meetings a part of the contract administration process, whether they consist of phone calls or formal meetings.

Consistency and an appropriate agenda are really important. Chapter 3 outlines a three-part agenda that provides a good basis for managing such meetings. In the first part, we keep the prewritten minutes that we simply adapt if they turn out to be wrong. This part of the meeting is to record the status of critical points to the contracting parties. Here we note that nobody has delayed anyone else, that payments are up to date and so on. The existence of these "standard" minutes has helped in the past by encouraging solution of a problem before the meeting gets going.

Such a solution will go into the second part of the meeting and will be followed by the rest of the standard agenda, which focuses on the upcoming plans and progress, rather than the past. The third part of the meeting is for recognition of performance and any outstanding contributions from members of the project team. A balanced set of minutes that regularly includes formal recognition of contributions will encourage people to read them, as well as help maintain good relationships between all concerned.

Use a consistent three-part agenda for progress meetings.

If we can handle as many issues as possible before they are brought up in the regular meeting, we will help keep them at the working level where they stand a better chance of being resolved. Part of the success that companies have had in this approach rests with having minimal bureaucracy, as well as delegating authority for solving problems to the lowest possible level in the organization. The closer the decision-maker is to the coal face, the faster and probably the more effective the solution will be. At this level, people from the buyer and seller organizations can trade favors, solve problems and build a team that will deliver a better contract outcome than their managers can, as the latter are often too remote from the details.

Try to simplify all of the processes involved in managing your contracts. Clients and contractors alike prefer to avoid unneeded bureaucracy.

Why we still need people for contracts administration.

The less formal the procedures we end up with, the more effective the day-to-day operation will be. But we may well lose the necessary history of decisions and who did what as the paper trail thins out. For this reason, and to ensure that the right people know what is going on, all decisions should be documented in some way. My preferred approach is to use a four-part carbonless form, 8 ½" × 11" or European A4 size. The size makes it easier to file and harder to lose. The top copy serves as confirmation to the person who received a verbal instruction or requested a decision. This allows the recipient to accept or challenge the communication. The second copy is sent to the contract file. The originator keeps the third copy, and the fourth copy is a spare should anyone need it. On this form we record the date and time of the discussion, who was present, and the details of what was agreed.

We need to document what we agree on as we go. Basic documentation serves as an aid to communication and provides a safety net in case things go wrong.

12.9 Thou Shalt deal Rationally and Fairly with Inevitable Changes

> *Change orders are a thorn in many people's sides. They can be managed more effectively if the ground rules are set well.*

Every contract I have ever worked on has had something change. In some cases, the change was small. In others, the very nature of the work to be done had changed. In between, we saw small and large changes to specifications, sequencing of work, priorities, and more. These changes are normal. Today, as we try to respond to a business and technical environment that leaves us bombarded by change, we need to be even more flexible and accommodating in how we deal with such things. Specifically, the need for our contracts to accommodate change is more critical today than it has ever been.

Change is inevitable. Be sure the contract can handle changes in the context of the project, as well as in the business environment that is defined by the companies that are participants.

As with everything else, we need to make the process of managing changes to a contract as simple as possible. Furthermore, we need to manage changes in a consistent way, so that those involved can gain confidence in the process and can therefore rely on our process to work under all conditions. Finally, we need to have a process that is fast. Speed usually means everything is kept simple and authority has been delegated to the lowest possible level consistent with the policy of each organization.

The change order process should be as simple, consistent and fast as possible.

Chapter 9 shows how the markup for a change is normally set by the owner. I think it makes sense to ask the contractor what it wants as a markup. Then adjust the price of the bid by that additional cost based on a predetermined value of changes. This encourages two things to happen: First, the contractor needs to balance competitiveness with opportunism in setting the markup rates for profit and overhead, as the rates will affect the competitiveness of the overall bid. And second, if we have also qualified exactly what is to be covered by these charges, then we have introduced a degree of consistency that is so important to the change management process later.

On larger contracts, or when the range of changes may be considerable, it may make sense to have more than one markup rate, depending on the nature of the change. For example, any change that costs less than $1,000 may have a fixed change fee of $100 to cover overheads and administration. After that, the rate may be 10% for the next $19,000, and 5% on any amount over that. A similar scheme may be used for profit rates. And for profit, we may wish to use different rates to be applied to direct work or supply versus subcontracted components that may carry a lower figure.

Make sure that appropriate markups are part of the change pricing structure.

Another perennial problem with change orders is an assessment of the effect of changes on the overall project for the contractor. Contractors are understandably reluctant to commit to the effect of one or more changes until they can assess the situation in the clear light of hindsight. This often means waiting until the end of a contract, then presenting the client with a huge bill. Equally understandable is the client's reaction to an added cost of indeterminate size that cannot be managed or processed until the end of the project when most of the opportunities to explain or absorb the cost have disappeared.

The SMART solution is to agree to a set of points in the life of the contract at which these adjustments are made for changes to that point in time. These points are usually milestones defined in terms of a set of deliverables that will be produced by that time and a target date. If the deliverables have been produced on or before the target date, then the effect will be small or zero, unless significant effort was expended in making good the schedule. (The Priority Triangle can help determine whether this should be done or not.) If the target date was missed, then we need to determine how much of that delay was attributable to changes and how much was attributable to other reasons. Here the regular weekly progress meeting minutes should help. By breaking up the effects of change into discrete pieces like this, we reduce the shock factor and can make better and more informed decisions as we go. The contractor should be involved in setting the dates for these adjustments, as this will help achieve buy-in to the process.

Add key points in the plan for delivery of the goods or services when overall adjustments to reflect the effects of changes can be made.

Once we have a change, we need to agree that it really is a change to the contract, and then we need to agree to the value of the change. If the vendor asks for a change order for something that the buyer feels is within the scope and intent of the contract, then we are faced with a dispute. So we need to do what we can to make any change readily discernible as such. Chapter 9 explains how the Work Breakdown Structure (WBS) or a SMART Breakdown Structure (SBS) would help us achieve this baseline or yardstick that we could identify changes against.

> **The key to effective management of changes is to be able to recognize them and assess them when they occur.**

There are several causes for changes to a contract. These include imperfections (errors, omissions) in the contract documents. These are just about inevitable given the fees or internal costs that buyers are willing to absorb to prepare a set of contract documents and the ever-shortening time that seems to be available for this work. Whatever the cause, the costs associated with these changes are for the buyer to absorb. An appropriate contingency or budget item should be allowed for this. The amount in that budget will be a reflection of the care that could be put into the preparation before the contract is signed. In my experience, many of the items that fall into this category are ones that the buyer would have had to pay for to get the desired outcome anyhow. The money has just been moved from the original contract (had there been enough time to add the missing detail) to the change order.

The other two main areas for change are those requested by the buyer and those requested by the seller. Buyers are allowed to change their minds, and doing so at a later point simply triggers a change that the buyer then pays for. Changes requested by the vendor are the result of better ideas, opportunities to take cost out of the project, or something missed in the bid. For the first reason, the buyer may benefit from the better idea. If there is cost to be taken out of the project, depending on the project priorities, we can see a benefit for both the buyer and the seller if the savings are shared and the rest of the project objectives are not compromised. Finally, if there is a change requested because of an error in the bid, then maybe the contractor is liable for its own mistake. But if a buyer can make a mistake in the several months it had to prepare the bid documents, can we forgive a contractor who makes a mistake in the mere days or weeks it has to prepare the bid? This may, under the right circumstances, be an opportunity to build a relationship with that contractor.

We need to recognize that changes are the result of imperfections in contract documents, client- or vendor-requested changes, or differing circumstances to those contemplated by the contract. Then we need to take responsibility for those that we instigated.

12.10 Thou Shalt Resolve Disputes in the most Equitable and Efficient Way Possible

> *Avoid the avoidable disputes, and resolve the*
> *rest as quickly and efficiently as possible.*

This part of Chapter 12 is largely unnecessary if we rationally apply all we have learned from the first nine sections. Failing that, we are still likely to have the occasional dispute. History and studies have repeatedly told us that the faster we resolve contractual disagreements, the more likely we are to have a reasonable outcome. Equally, we know that it is best to get people involved in the real work and familiar with the details and the history to address and – ideally – resolve the problem.

The best way to manage a dispute is to resolve it as quickly as possible and at the lowest level in the organization that we can.

The mechanisms we have already suggested to minimize the incidence of disputes include identifying latent disputes, making sure we manage each other's – and our stakeholders' – expectations, and having regular meetings to facilitate this. We have also seen the value of good documentation, such as a paper trail of all decisions and a well-kept diary. Chapter 10 identifies alternative dispute resolution options. Here is one more. It is aimed at heading any dispute off at the pass.

The best contracts managers encourage participants to solve problems as they come up, and to do so quickly. We want to harness this by formalizing it. The following steps help us reduce, and in some instances even eliminate, the incidence of disputes:

1. Elect a person from each organization at the working level (perhaps the project manager, a superintendent on a site and his client counterpart or a contracts administrator) to be the prime contact for the contract or a specific part of a larger contract.

2. These two people should be delegated as much authority to resolve problems and commit their organizations to their decisions as is both reasonable and possible. Some companies and organizations have policies that dictate limits on spending, so this should be respected.

3. Give these two people 48 hours to resolve any dispute before it is automatically escalated to the next level.

4. The people who will address the issue at the next level will have greater authority and spending limits. They, too, have 48 hours to resolve the problem, and they should rely on the advice and guidance of the people who have brought the problem to them.

5. Only if the problem persists at this level is it automatically escalated again. This time it is brought to the weekly progress meeting. If still unresolved, then it is escalated again.

6. There are several ways that any remaining problems at this level can be addressed. Consider the alternative dispute resolutions options discussed in Chapter 10. Alternatively, consider using a third-party assessor who makes a temporary decision that is binding until the end of the project so that the work can continue. The problem, if it has not gone away by then, can be resolved without time pressures.

When I have seen this approach used, the results have been consistently good. Very few problems ever get escalated even once. I have yet to see anything that needs to have Step 6 applied.

 Have a mechanism in place to avoid issues becoming disputes.

Use the weekly progress meeting not just for its original purpose, to resolve any disputes that have not been resolved in 96 hours, but also to monitor the mood of the players in the contract. Again, the tool has been discussed before, but is worth looking at a bit differently. Chapter 8 suggests having a kick-off meeting at which the contractor is brought on board with the rest of the project team. At this meeting, the project charter is normally updated. Also at this meeting, we should ask each other what we did that annoyed others and what others did that annoyed us. We use this as the basis for developing part of the fixed agenda for each weekly progress meeting. We can use these regular assessments to gauge the feelings of the team. If we have face-to-face meetings, we can watch body language, listen hard to people who may be trying to deliver subtle messages, and use whatever other tricks we have in order to detect changes in the mood of individuals, companies, or the entire group.

Why bother with this? The emotions of the participants often provide early warning signs of festering problems that have not been picked up or eliminated by all of the other tools and processes we have deployed. Sometimes it pays to have a neutral third party doing this monitoring, as such a person will have no bias toward either cause and will therefore be more objective.

Monitor people's concerns and moods as the project progresses. Do not let side issues take over and sour the contract.

Disappointment is the result of not having our expectations met. Sometimes this is because we failed to communicate those expectations. Sometimes it is because there was no intention to meet them. We need to work on communication and honest management of each other's expectations throughout the life of the contract.

Manage expectations and relationships. They are the key to successful trading.

The end result of all of this effort? We have one of the best contracts.

The best contract has no disputes.

12.11 Closure

This book has identified 10 commandments of better contracting. We ignore these commandments at our peril. The best buyers and sellers follow many of these but rarely all ten of the commandments are truly adopted.

Experienced practitioners will know what makes sense and why. Unfortunately, once they are sufficiently experienced, they retire and we end up perpetuating problems that have been inherent (and arguably abhorrent) for a long time.

Let me close with the positive benefits we derive from each of the commandments and the synergy between them.

If we follow the first commandment we end up with simpler, easier-to-understand and more equitable contract documents, simpler processes and the potential for greater cooperation through higher levels of trust.

If we follow the second commandment, we will encourage vendors to work within their competence and take fewer risks that lead to failure in some way for either party.

The third commandment tells us to listen to our client. Even buyers have clients. We need to really understand what our customers want. Only then can we deliver to their satisfaction. If we get this right, we reduce changes, rework, disappointment and all the associated grief, mistrust, cost and delays.

The fourth commandment admonishes us to select our vendors based on **value,** not **lowest price**. Cheap is cheap. It is often not the best. Even the apparently lowest initial price is no guarantee that we will end up with the lowest final cost. Selecting on value forces us to consider contractors and their capabilities using a more balanced score card.

Following the firth commandment helps us avoid misunderstandings that result from ambiguity. This will also help build trust as there will be more open and complete communication as well as fewer disagreements later.

The sixth commandment admonishes us to be business-like in risk assignment. This reduces risk premiums, subsequent arguments and eventually leads to fairer contracts and greater trust. We cannot price the last item (trust) but we do know that higher trust equates to lower cost.

The seventh commandment addresses trust but cautions us to be wise leaders in establishing a trust-based working relationship.

The eighth commandment addresses the need to maintain trust and open communication by not messing around with the terms, conditions and intent of the contract once we have signed it.

The ninth commandment recognizes that change is inevitable, and that we need to embrace and manage good changes, reject inappropriate ones and compensate contractors for the added effort required in addressing them.   If changes lead to cost savings; these should be equitably dispensed.

The final commandment advocates an intelligent approach to dispute resolution. This includes avoiding emotional responses, being fair to our business partners and being prepared to compromise when this leads to a simpler and more business-like solution.

Put all of these commandments to work. Encourage your business partners to develop a shared vision with you about how to do business more effectively.   Build on your own successes and experience by harnessing the knowledge and wisdom of others – and by sticking to the Ten Commandments of Better Contracting.   Doing so will make the experience, the process and the deal more rewarding for all.

REFERENCES AND BIBLIOGRAPHY

Chapter 1   Thou Shalt Contract Within the Law and the Working Environment of the Contracting Parties

American General Contractors. (1996). *Project Delivery Systems*.

American Institute of Architects. (1991). Document A121/CM: Standard of Agreement Between Owner and Construction Manager, Who Is the Contractor. Washington, DC: AIA.

Barnes, M. (1988). "Construction Project Management." *International Journal of Project Management*, 5, 69-79.

Boyce, Tim. (1991). "The Commercial Engineer". Hawksmere, London, UK.

Clough, R. H. and Sears, G.A. (1994). "Construction Contracting", 6th Edition, John Wiley & Sons.

Gilbreath, Robert D. (1992). "Managing Construction Contracts: Operational Control for Commercial Risks" Second edition, John Wiley and Sons, U.S.A.

Morris, P.W.G. Editor. (1997). "Management of Projects." *American Society of Civil Engineers* (Thomas Telford, Ltd.).

Wright, J. and Hughs, W. "Construction Contracts Law and Management".

Chapter 2   Thou Shalt Not Mix Up the Wrong Work Packages

Associated General Contractors of America. AGC Documents 400, 410.

Associated General Contractors of America. (1997). AGC Document 510: Standard Form of Agreement Owner and Construction Manager (With CM as Owner's Agent). Washington, DC: AGC.

Construction Industry Institute (CII). Publications:
RS6-4 Contractor Planning for Fixed-Contracts. (1987)
RS5-1 Impacts of Various Construction Contract Types and Clauses on Project Performance. (1986).
RS 111-1 Owner/Contractor Work Structure: A Preview,
IR7-3 Procurement and Materials Management, A Guide to effective Project Execution. (1999).
SD-20 Construction Contractor Planning for Fixed-Price Construction. (1986).

SD-10 Determining the Impact of various Construction Contract Types and Clauses on Project Performance, Vol. 1: Analysis and Recommendations. (1986).

SD-11 Determining the Impact of various Construction Contract Types and Clauses on Project Performance, V. 11, Appendices, (1986).

RS130-1 Contractor Supplier Relationships: A Project Delivery System to Optimize Supplier Roles in EP Projects. (1998).

RR130-11 PEpC: A Breakthrough Project Delivery System that Improves Performance by Reforming Owner, Contractor Supplier Relationships.

IR7-3 Procurement and Materials Management: A Guide to Effective Project Execution. (1999).

Construction Management Association of America. (1993). CMAA Documents A-1, A-2, A-3 & A-4: Standard Forms of Agreement Between Owner and CM, Owner and Contractor, General Conditions of Contract Between Owner and Contractor, and Owner and Design Professional. CMAA.

Loftus, J., (Editor). (1999). "Project Management of Multiple Projects and Contracts", American Society of Civil Engineers (Thomas Telford, Ltd.)

O'Reilly, M., (1996). "Civil Engineering Construction Contracts", American Society of Civil Engineers (Thomas Telford, Ltd.).

Park, W. (1979). "Construction bidding for profit" 168-177. New York: John Wiley & Sons.

Chapter 3   Thou Shalt Listen to, and Understand, the Real Wishes and Needs of the Customer

Al-Harbi, K.M.A. (1998). "Sharing fractions in cost-plus-incentive-fee contracts." *International Journal of Project Management*, Vol. 16, No. 2, 1998, 73-80.

Alhazmi, T. and McCaffer, R. (2000). "Project Procurement System Selection Model" *Journal of Construction Engineering and Management*, May/June, 176-184.

American Institute of Architects. (1987). Document A201: The General Conditions of the Contract for Construction. Washington, DC; AIA.

Arditi, D. and Yasamis, F. (1998). "Incentives/Disincentive Contracts: Perceptions of Owners and Contractors", *Journal of Construction Engineering and Management*, Vol. 124, No. 5, September/October, 361-373.

Ashley D.D. and Bonner, J.J. (1987). "Political risk in international construction." *Journal of Construction Engineering and Management*, 113(3), 447-467.

Ashworth, A. (1991). "Contractual Procedures in the Construction Industry (2nd ed.)", Longman Scientific and Technical Group, Harlow, UK.

Bainbridge, L. (1996). "The Partnering Process. Partnering in Design and Construction." McGraw-Hill, 89-109.

Bajaj, D., Oluwoye, J., & Lenard, D. (1997). "An analysis of contractors' approaches to risk identification in New South Wales, Australia." *Construction Management and Economics*, IS, 363-369.

Baronm, D.P. (1972). "Incentive contracts and competitive bidding." *The American Economic Review*. 62, p 384-394.

Bassok, Y. and Anupindi, R. (1994). "Analysis of supply contracts with total minimum commitment." Working Paper, Northwestern University, Evanston, IL 60208.

Batavia, R. (2001). "How to Maximize Project Success with the Right Contracting Strategy." *Proceedings of the PMI Annual Seminars and Symposiums,* September 7-16, Houston, Texas, USA.

Bentil, K. (1989). "Fundamentals of the Construction Process." Kingston, MA: R.S. Means.

Berends, T.C. (2000). "Cost plus incentive fee contracting – experiences and structuring." *International Journal of Project Management*, Vol. 18, No: 3, 2000, 165-171.

Beveglia-Zampetti, A. (1997). "The UNCITRAL Model Law on Procurement of Goods, Construction and Services." In Hoekman and Mavroidis, Eds.

Black, C., Akintoye, A., and Fitzgerald, E. (2000). "An analysis of success factors and benefits of partnering in construction." *International Journal of Project Management*, Vol. 18, No. 6, 423-434.

Boyce, Tim. (1991). "The Commercial Engineer". Hawksmere, London, UK.

Branco, F. (1994). "Favoring Domestic Firms in Procurement Contracts" *Journal of International Economics*, 37:65-80.

Breton, A., and Salmon, P. (1995). "Are Discriminatory Procurement Policies Motivated by Protectionism?" Kyklos, 49:47-68.

Carr, R. I. (1997). "Paying the price for construction risk", *Journal of Construction Division*, ASCE, 203(CO1), 153-161.

Caron, F., Marchet, G., and Perego, A. (1998). "Project logistics: integrating the procurement and construction processes", *International Journal of Project Management*, Vol. 16, No.5. 311-310.

Cavendish, P. and Martin, M. D. (1987), "Negotiating and Contracting for Project Management". The Project Management Institute.

Chapman, C.B., Ward, S.C., and Bennel, J.A. (2000). "Incorporating uncertainty in competitive bidding", *International Journal of Project Management*, Vol. 18, Issue 5, 2000, 337-347.

Chen, X. (1995). "Directing Government Procurement as an Incentive of Production", Journal of Economic Integration 10:130-40.

Clough, R. Construction Contracting. John Wiley and Sons, New York.

Cohen, M.A, and Agrawal, N. (1999). "An analytical comparison of long and short term contracts." *IIE Transactions Norcross*, August, Vol. 31, Issue: 8, Pages: 783-796.

Collier, K. (1987). "Construction Contracts." Englewood Cliffs, NJ: Prentice-Hall.

Construction Industry Institute (CII). Publications:

RS6-4 Contractor Planning for Fixed-Contracts. (1987).

RS5-3 Contract Risk Allocation and Cost Effectiveness. (1988).

RS5-1 Impacts of Various Construction Contract Types and Clauses on Project Performance. (1986).

RS5-2 Incentives Plans: Design & Application Considerations.

RS 111-1 Owner/Contractor Work Structure: A Preview.

IR7-3 Procurement and Materials Management, A Guide to effective Project Execution. (1999).

SD-20 Construction Contractor Planning for Fixed-Price Construction. (1986).

SD-10 Determining the Impact of various Construction Contract Types and Clauses on Project Performance, Vol1: Analysis and Recommendations. (1986).

SD-11 Determining the Impact of various Construction Contract Types and Clauses on Project Performance, Vol. 1, Appendices, (1986).

RS-114-1 Innovative Contractor Compensation. (1998).

RR114-11 Innovative Strategies for Contractor Compensation, (1998).

RS130-1 Contractor Supplier Relationships: A Project delivery system to Optimize Supplier Roles in EP Projects. (1998).

RR130-11 PEpC: A Breakthrough Project Delivery System that Improves Performance by Reforming Owner, Contractor Supplier Relationships.

IR7-3 Procurement and Materials Management: A Guide to Effective Project Execution (1999).

Construction Industry Research and Information Association Tunneling. (1977). "Improved Contract Practices." UK.

Cook, L. (1990). "Partnering: contracting for the future", *ASCE Journal of Construction Management and Engineering*, 6(4), 431-446.

Cox, A. and Thompson, I. (1998). "Contracting for Business Success." American Society of Civil Engineers (Thomas Telford, Ltd.)

Cox, A. and Townsend, M. (1998). "Strategic Procurement in Construction." Thomas Telford, London.

Cruz, J. C. (1998). "Evaluation of Project Delivery Systems." Graduate report. University of Florida.

Debroy, B., and Pursell, G. (1997). "Government Procurement Policies in India." In Hoekman and Mavroidis, Eds.

Deltas, G., and Evenett, S. (1997). "Quantitative Estimates of the Effects of Preference Policies." In Hoekman and Mavroidis, eds.

Design/Build Contractors Offered Coverage. *Building Design & Construction*. (1996). September, 12.

Dissanayaka, S.M. and Kumaraswany, M.M. (1997). "Re-engineering Procurement Frameworks and Relationships." *Proceedings of the International Conference on Construction Process, Re-engineering*, Gold Coast, Australia, 157-167.

Dombkins, D. (1990). "Partnering: A definition, review of American and Australian experience, theoretical framework and recommendations for implementation in Australia." School of Building, University of New South Wales.

Domberger, S., Hall, C. and Lee, E.A.L. (1995). "The Determinants of Price and Quality in Competitively Tendered Contracts." *Economic Journal*, 105:1454-70.

Dozzi, P., Hartman, F., Tidsbury, N. and Ashrafi, R. (1996). "More-Stable Owner-Contractor Relationships." *ASCE Journal of Construction Engineering and Management*, March, Vol. 122, No. 1, 30-35.

General Services Administration. (1990). *Construction Management Guide*. Washington, DC: GSA.

Fellows, R.F.F. (1997). "The Culture of Partnering." *Proceedings of the CIB W-92 Procurement Systems Symposium, Procurement-A key to Innovation*, University of Montreal, 21-25 May, 193-202.

Franks, J. (1990). "Building Procurement Systems: A Guide to building project management" (2nd ed.). Chartered Institute of Building, Ascot, UK.

Fried, C. (1981). "Contract as Promise, A Theory of Contractual Obligations." Harvard Press, Cambridge, Massachusetts.

Gaafar, H.K. and Perry, J.G. (1999). "Limitation of Design Liability for Contractors." *International Journal of Project Management*, Vol. 17, No. 5, 301-308.

Gordon, C.M. (1994). "Choosing Appropriate Construction Contracting method", *Journal of Construction Engineering and Management*, Vol. 120, No.1, March, 196-210.

Hart, O., and Holmstream, B. (1987): "The Theory of Contracts," in *Advances in Economic Theory*, 5th World Congress of the Econometric Society. Cambridge: Cambridge University Press.

Hartman, F.T. (1993). "Better Construction Contracts, the Secret Ingredient." *Proceedings of the Project Management Institute (PMI). Northwest Regional Symposium*, Calgary, 708-715.

Hartman, F.T. (1994). "Contracts don't work: so what's the alternative?" *Proceedings of the 25th International Symposium of the Project Management Institute*, Vancouver, October, 178-183.

Hartman, F.T. (1995). "Re-engineering the Construction Contract." *Proceedings of the First International Conference on Construction Project Management*, Singapore, 47-58.

Hawwash, K.I.M. and Perry, J.G. (1996). "Contract Type Selector (CTS): a KBS for training young engineers." *International Journal of Project Management*, Vol. 14, No. 2, 95-102.

Hermalin, B. (1988): "Three Essays on the Theory of Contracts." Ph.D. Thesis. Massachusetts Institute of Technology.

Hoekman, B. and Mavroidis, P.C. (eds). (1997). "Law and Policy in Public Purchasing: The WTO Agreement on Government Procurement." Ann Arbor: University of Michigan Press.

Holt, G.D., Olomolaiye, P.O. and Harris, F.C. (1993). "Tendering practice – exploring alternatives", *Faculty of Building Journal*, Autumn, 28-30.

Holt, G.D., Olomolaiye, P.O. and Harris, F.C. (1993). "A conceptual alternative to current tendering practice." *Building Research and Information*, 21(3), 167-172.

Hughes, W.P. (1992). "Identifying Appropriate Construction Procurement Strategies." Paper at 5th Annual Construction Law Conference, September. King's College, London, UK.

Huston, C. L. (1998). "The ABCs of DPC: A primer on Design-Procurement-Construction for the Project Manager." Project Management Institute.

Institution of Civil Engineers. (1993). "ICE Conditions of Contract." American Society of Civil Engineers (Thomas Telford, Ltd.)

Kaufman, D. (1997). "The Future of Design-Build." Graduate report. University of Florida.

Kitchens, M. (1996). "Estimating and Project Management for Building Contractors." American Society of Civil Engineers (ASCE Press).

Lanford, D.A. and V.R. Rowland, (1995). "Managing Overseas Construction Contracting." American Society of Civil Engineers (Thomas Telford, Ltd.)

Larson, Erik W., and John A. Drexler, Jr. (1997). "Barriers to Project Partnering: Report from the Firing Line." *Project Management Journal*, (March) V. 28, No. 1: 46–52.

Latham, M. (1994). "Constructing the Team: Joint review of Procurement and Contractual Arrangements in the UK Construction Industry." Department of the Environment, UK.

Lenk, B.R. (1977). "Government procurement policy: A survey of strategies and techniques." Office of Naval Research, Virginia, Final Report.

Love, P.E.D., Gunasekaran, A. and Li, H. (1998). "Concurrent engineering: a strategy for procuring construction projects." *International Journal of Project Management*, Vol. 16, No.6, 375-383.

Lusch, R. F. and Brown, J.R. (1996). "Interdependency, Contracting, and Relational Behavior in Marketing Channels." *Journal of Marketing*, 60 (October): 19-38.

Macaulay, S. (1963). "Non-Contractual Relations in Business: A Preliminary Study." *American Sociological Review*, 28 (February), 55-67.

Marsh, P.D.V. "Contracting for Engineering and Construction Projects." Gower Press.

Marshall, V. (1997). "Design/Build: The Project Delivery System of Choice." Graduate report. University of Florida.

Masterman, J.W.E. (1992). "An Introduction to Building Procurement Systems", E&FN Spon Ltd, London.

McCall, J.J. (1970). "The simple economics of incentive contracting", *The American Economics Review*, 60:837-864.

McCarthy, S.C. and Tiong, R.I.K. (1991). "Financial and contractual aspects of build-operate-transfer projects" *International Journal of Project Management*, 9:222.

McVay, B. (1995). "Caution! Slippery Fixed-Price Incentive Contracts Ahead!" *National Contract Management Journal* 26 (1):15-24.

Merna, A. and Smith, N.J. (1990). "Bid evaluation for UK public sector construction contracts." *Proceedings, Institution of Civil Engineers*, Part 1(88), 91-105.

Moore, C., Mosley, D. and Slagle, M. (1992). "Partnering: Guidelines for Win-Win Project Management." *Project Management Journal* 23, No. 1 (March): 18–21.

Naoum, S.G. and Langford, D. (1987). "Management Contracting: the client's view." *Journal of Construction Engineering and Management*, 113(3), 369-384.

NWPC and NBCC. (1990). "No Dispute-Strategies for Improvement in the Australian Building and Construction Industry." *A report by the National Public Works Conference and National Building and Construction Council Joint Working Party*, May, Canberra, ACT, Australia.

O'Reilly, M., (1996). "Civil Engineering Construction Contracts", American Society of Civil Engineers (Thomas Telford, Ltd.).

Park, W. (1979). "Construction bidding for profit" New York: John Wiley & Sons. 168-177.

Quartey, EL Jnr. (1996). "Development projects through build-operate schemes: their role and place in developing countries" *International Journal of Project Management*, Vol. 15, No. 1, 47-52.

Rothenberg, J. (1993). "Comment." In Jim Leitzel and Jean Tirole, (Eds.), *Incentives in Procurement Contracting*. Boulder: Westview Press.

Runeson, K.G. and Skitmore, R.M, (1999). Tendering theory revised." *Constr. Mgmt. and Economics*, 17(3), p.285-296.

Russell, J.S. (2000). "Surety Bonds for Construction Contracts", American Society of Civil Engineers (Thomas Telford, Ltd.).

Skitmore, M., Drew, D. and Ngai, S. (2001). "Bid Spread." *Journal of Construction Engineering and Management*, Vol. 127, March/April, 149-153.

Stinchcombe, A.L. (1985), "Contracts as Hierarchical Documents," Organization Theory and Project Management Administering Uncertainty in *Norwegian Offshore Oil*.

Sweetman, K. (1996). "Procurement", *Harvard Business Review*, (74), 11-13.

Schwartz, A. (1992): "Legal Contract Theories and Incomplete Contracts." in *Contract Economics*, ed. by L. Werin and H. Wijkander. Oxford: Basil Blackwell, 6, 76-108.

The Institution of Civil Engineers (1993). "The New Engineering Contract", Thomas Telford Ltd., London.

The Institution of Civil Engineers (1995). "The Engineering and Construction Contract", Thomas Telford Ltd., London.

Thompson, I. and Cox, A. (1998). "Contracting for Business Success", American Society of Civil Engineers (Thomas Telford, Ltd.).

Tiong, R.L.K. (1992). "Strategies in Risk Management of On-Demand Guarantees." *Journal of Construction Engineering and Management*, Vol. 118, No. 2, June, 229-243.

Tsay, A.A. and Lovejoy, W.S. (1998). "Quantity flexibility contracts and supply chain performance". Working Paper, Graduate School of Business, Stanford University, Stanford, CA.

Turner, A. (1990). "Building Procurement", McMillan, Houndsmill, UK.

Ward, S. and Chapman, C. (1994). "Choosing contractor payment terms" *International Journal of Project Management*, Vol. 12, 216.

Wood, D. (1997). "The WTO Agreement on Government Procurement: An Antitrust Perspective". In Hoekman and Mavroidis, Eds.

Woodward, J. (1997). "Construction Project Management: Getting It Right the First Time", American Society of Civil Engineers (Thomas Telford, Ltd.)

Woosley, P. (1981). "International Contracting" Construction Litigation, eds. K.M. Cushman, B.W. Ficken and W.R. Sneed, Practicing Law Institute, New York, 753-787.

Wright, J. and Hughs, W. "Construction Contracts Law and Management".

Chapter 4   Thou Shalt Not Blindly Pick the Contractor who was Cheapest Because It may have Made the Biggest Mistake

Ahlbrant, R. S., Jr. (1974). "Implications of contracting for public service." *Urban Affairs Quarterly*, 9:337-358.

Alsugair, A.M. (1999). "Framework for evaluating bids of construction contractors", *ASCE Journal of Management in Engineering*, 15(2):72-78.

Arnavas, D., and Ruberry, W., (1994). "Government Contract Guidebook", Second Edition. Washington, DC: Federal Publications.

Barnes, M. (1988). "Construction project management." *International Journal of Project Management*, 5:69-79.

Beale, H. and Dugdale, T. (1975). "Contract Between Businessmen: Planning and the Use of Contractual Remedies", *British Journal of Law and Society*, 2 (1):45-60.

BW 10/96a. (1996). "Preston Gives DOD Acquisition Scorecard", *Contract Management*, Vol. 36, No. 10 (October), 36-38.

BW 10/96b. (1996). "DOD to Tailor Past Performance to Business Requirements", *Contract Management*, Vol. 36, No. 10 (October), 38.

BW 6/97. (1997). Final Edition, *Contract Management*. Vol. 37, No. 6 (June), 31.

Chua, D.K.H. and Li, D. (2000). "Key Factors in Bid Reasoning" *Journal of Construction Engineering and Management*, 126:349-357.

Chua, D.K.H.; Li, D.Z. and Chan, W.T. (2001). "Case-Based Reasoning Approach in Bid Decision Making"." *Journal of Construction Engineering and Management*, 127:35-45.

C.I.C. (1994). "The Procurement of Professional Services, Guidelines for the Value Assessment of Competitive Tenders." Construction Industry Council, UK.

Contract Management. (1997). "OFPP Delays Past Performance Requirements for Contracts Under $1 Million", Final Edition, Vol. 37, No. 2 (February), p. 31.

Collier, Keith, (1979). "Construction Contracts", Reston Publishing Company, Inc.

Corcoran, J. and McLean, F. (1998). "The selection of management consultants-How are governments dealing with this difficult decision? An exploratory study." *Int. Journal Public Sector Mgmt.*, Bradford, UK. 11(1), 37-54.

Crowley, L.G. and Hancher, D.E. (1995). "Evaluation of Competitive Bids" *Journal of Construction Engineering and Management*, Vol. 121, June, 238-245.

Crawley, L.G. and Hancher D.E. (1995). "Risk Assessment of Competitive Procurement." *Journal of Construction Engineering and Management*, Vol. 121, June, 230-237.

DeHoog, R. H. (1990). "Competition, negotiation, or cooperation: Three models for service contracting." *Administration and Society*, 22, Nov., 317-340.

Docherty, J., and D. Langford. (1993). "Structure, Strategy and Survival in Scottish Construction Firms." *Proceedings of ARCOM*. Cambridge.

Domberger, S., Hall, C., and Li, E. (1994). "The determinants of quality in competitively tendered contracts", (Working Paper Series). Sydney, Australia: Graduate School of Business, University of Sydney.

Domberger, S., Meadowcroft, S. A., and Thompson, D. J. (1986). "Competitive tendering and efficiency: The case of refuse collection." Fiscal Studies, 7(4), 69-87.

Domberger, S.. Meadowcroft, S., and Thompson, D. (1987). "The impact of competitive tendering on the costs of hospital domestic services." Fiscal Studies, 8(4), 39-54.

Domberger, S., Hall, C., and Lee, A.H.L. (1995). "The Determinants of Price and Quality in Competitively Tendered Contracts." *Economic Journal,* 105:1454-70.

Drew, D.S., and Skitmore, R.M. (1990). "Analyzing Bidding Performance; Measuring the Influence of Contract Size and Type." *Building Economics and Construction Management: Management of the Building Firm*. Sydney, Australia: CIB-W-65.

Drew, D. (1995). "The effects of contract type and size on competitiveness in construction contract bidding" Ph.D. Thesis, University of Salford, Salford, England.

Drew, D.S., and Skitmore, R.M. (1997). "The Effect of Contract Type and Size on Competitiveness in Bidding." *Construction Management and Economics*, 15 (1997): 469-489.

Drexler, J.A. Jr. and Larson, E.W. (2000). "Partnering: Why Project Owner-Contractor Relationships Change, *J. Constr. Engrg. and Mgmt.* 126, Jul/Aug, p.293-297.

Egginton, B. (1996). "Multi-national consortium based projects: improving the process" *International Journal of Project Management*, Vol. 14, No. 3, 169-172.

Fayek, A. (1998). "Competitive Bidding Strategy Model and Software System for Bid Preparation" *J. Constr. Engrg. and Mgmt.* Vol. 124, Jan/Feb. 1-10.

FCR 5-12-97c. (1997). "Decisions in Brief: Past Experience -- Service Contracts," Federal Contracts Report, Vol. 67, (May 12). 570-571.

Ferguson, N.S., Langford, D.A. and Chan, W.M. (1995). "Empirical study of tendering practice of Dutch municipalities for the procurement of civil-engineering contracts." *International Journal of Project Management*, Vol.13, No 3, 157-161.

Friedman, L., A. (1956). "Competitive Bidding Strategy." *Operations Research*, Vol. 4, February, 104-112.

Fuerst, M.J., (1977). "Theory for Competitive Bidding." *Journal of the Construction Division*, Vol. 103, March, 139-152.

Gates, M., (1967). "Bidding Strategies and Probabilities", *Journal of the Construction Division*, Vol. 93. June 1967, 75-107.

GCS 15-96. (1996). "Proposal Evaluation: Protest Sustained Where Record Insufficient to Establish Equality Under Past Performance Criteria," *Government Contracts Service*, No. 15-96 (August 15).

GCS 16-96. (1996). "Past Performance: Where Past Performance Is More Important Than Price, Agency May Award on Price Alone Where Solicitation Stipulates Price Only May Be Considered for First-Time Offerors," *Government Contracts Service*, No. 16-96 (August 30).

GCS 18-96. (1996). "Past Performance: Past Performance Should Be Tailored to Particular Business Areas, Says DOD Study," *Government Contracts Service*, No. 18-96 (September 30).

GCS 21-96a. (1996). "Past Performance: Contractors Should Be Permitted to Refute Unfavorable Past Performance, Says Industry Group," *Government Contracts Service*, No. 21-96 (November 15).

GCS 21-96b. (1996). "Proposal Evaluation: Agency Failed to Consider Complexity of Contractor's Past Contracts," *Government Contracts Service*, No. 21-96 (December 15.

Globerman, S. and Vining, A.R., (1996). "A Framework for Evaluating the Government Contracting-Out Decision with an Application to Information Technology," *Public Administration Review*, Vol. 56, No. 6 (November/December). 577-586.

Gransberg, D.D., Dillon, W.D., Reynolds, L and Jack Boyd. (1999). "Quantitative Analysis of Partnered Project Performance", *Journal of Construction Engineering and Management*, Vol. 125, No.3, May/June, 161-166.

Griffith, F. (1989). "Project contract strategy for 1992 and beyond", *International Journal of Project Management*, 7:69-82.

Griffis, F.H. (1992). "Bidding Strategy: Winning over key Competitors." *Journal of Construction Engineering and Management*, Vol. 119, March, 151-165.

Hatush, Z. and Skitmore, M. (1997). "Criteria for contractor selection" *Constr. Mgmt and Economics*, London. 15(1), 19-38.

Hearn, E., "Federal Acquisition and Contract Management". Los Altos, CA.

Hensher, D. (1989). "Competitive tendering in the transportation sector", Economic Papers, 8(1), 1-11.

Hodge, G. A. (1996). "Contracting out government services: A review of international evidence." Melbourne, Australia: Monash University (Montech Pty. Ltd.).

Holt, G.D. (1995). "A methodology for predicting the performance of construction contractors." Unpublished Ph.D. thesis, University of Wolverhampton, UK.

Holt, G.D. (1998). "Which Contractor Selection methodology" *International Journal of Project Management*, Vol. 16, No. 3, 1998, 153-164.

Holt, G.D., Olomolaiye, P.O. and Harris, F.C. (1995). "Application of an alternative contractor selection model." *Building Research and Information*, 23(5), 255-264.

Holt, G.D., Olomolaiye, P.O. and Harris, F.C. (1995). "A review of contractor selection practices in the UK construction industry". *Building and Environment*, 30(4), 553-561.

Holt, G.D., Olomolaiye, P.O. and Harris, F.C. (1994). "Evaluating pre-qualification criteria in contractor selection." *Building and Environment*, 29(4), 437-448.

Holt, G.D., Olomolaiye, P.O. and Harris, F.C. (1994). "Evaluating performance potential in the selection of contractors." *Engineering Construction and Architectural Management*, 1(1), 29-50.

Holt, G.D., Olomolaiye, P.O. and Harris, F.C. (1994). "Factors influencing UK construction clients choice of contractor" *Building and Environment*, 29(2), 241-248.

Holt, G.D., Olomolaiye, P.O. and Harris, F.C. (1995). "Applying multi-attribute analysis to contractor selection decisions." *European Journal of Purchasing and Supply Management*, 1(3), 139-148.

Humphries, J. (1994). "Contractors understanding of the factors which affect tendering levels of construction works", M.Sc. dissertation, Dept. of Civil Engineering, Loughborough University of Technology, UK.

Jaafari, A. (1996). "Twinning Time and Cost in Incentive-Based Contracts', *ASCE Journal of Management in Engineering*, Vol. 112, No.4, July/August, 62-72.

Jaselskis, E.J. and Talukhaba, A. (1998). "Bidding Considerations in Developing Countries" *J. Constr. Engrg. and Mgmt*. Vol.124, May/June, 185-193.

Keer, P M., and Martin, L. L. (1993). "Performance, accountability, and purchase of service contracting." *Administration in Social Work*, 17:61-79.

Little, A.D., (1996). "Final Report for the Contractor Past Performance Systems Evaluation Study," June 17.

Martin, D. L., and Stein, R. M. (1993). "An empirical analysis of contracting out local government services." In G. Bowman, S. Hakim, & P Seidenstat (Eds.), *Privatizing the United States justice system: police, adjudication, and corrections services from the private sector* (82-106). Jefferson, NC: McFarland.

Martin, M.D., and Webster, EM. (1986). "Contract type and the measurement of project success." Project Management Institute Seminar & Symposium, Montreal, Canada.

McMillan, J, (1990). "Managing Suppliers: Incentive Systems in the Japanese and U.S. Industry", *California Management Review*, 32/4 (Summer): 38-55.

Mehay, S. L., and Gonzalez, R. A. (1993). "Direct and indirect benefits of intergovernmental contracting for police services." In G. Bowman, S. Hakim, & P Seidenstat (Eds.), *Privatizing the United States justice system: police, adjudication, and corrections services from the private sector* (67-81). Jefferson, NC: McFarland.

Merna, A. and Smith, N.J. (1990). "Project managers and use of turnkey contracts." *International Journal of Project Management*, 8:183 189.

Merror, E. and Yarossi, M.E. (1994). "Managing capital projects: Where have we been-Where are we going?" *Chemical Engineering*, 101, 108-111.

Molenaar, K.R. and Songer, A.D. (1998). "Model for public sector design/build project selection." *J. Constr. Engrg. and Mgmt.*, ASCE, 124(6), 467-479.

Morris, P.W.G. and Hough, G. H. (1987). "The Anatomy of Major Projects: A Study of the Reality of Project Management", Chichester, Wiley.

Musgrove, K.E. (1988). "A comparative study of district versus contracted pupil transportation systems in Missouri." Unpublished Ed.D. Dissertation, University of Missouri, Columbia.

Nadel, N.A., (1991). "Unit Pricing and Unbalanced Bids", *Civil Engineering*, Vol. 61, June 1991, 62-63.

Naphtine, R. and Smart, R. (1995). "Design and Build-lessons from the UK channel tunnel terminal", *Proceedings of the Institution of Civil Engineers, Civil Engineering*, 108, 123-130.

Palaneeswaran, E., Kumaraswamy, M.M. and Tam, P.W.M. (1999). "Comparing approaches to contractor selection for design and build projects." *Proc. Joint triennial Symp. (CIB W55, W65 and W92) on Customer satisfaction: A Focus for research and Practice*, P. Bowen and R.Hindle, eds. Vol.3, 936-945.

Peeters, W.A. (1996). "The Appropriate Use of Contract Types in Development Contracts." (A systems approach with emphasis on the European Space Sector) (Nordwij, the Netherlands, European Space Research and Technology Center).

Pinto, J.K., and Mantel, S.J. (1990). "The causes of project failure." *IEEE Transactions on Engineering Management*, 37 (4), 269-276.

Pinto J.K., and Slevin, D, P. (1988). "Project success: Definitions and measurement techniques", *Project Management Journal*, 19 (3), 67-73.

Prager, J. (1994). "Contracting out government services: Lessons from the private sector." *Public Administration Review*, 54, March/April, 176-184.

Russell, J.S. (1996). "Constructor Pre-qualification: Choosing the Best Constructor and Avoiding Constructor Failures", American Society of Civil Engineers (ASCE Press).

Russell, J.S., Hancher, D.E. and Skibniewski, M.J. (1992). "Contractor prequalifications data for construction owners", *Construction Management and Economics*, 10, 117-135.

Schleifer, T.C. (1990). "Construction Contractor's Survival Guide", John Wiley & Sons, New York, NY

Sivewright, K. (1996). "Transferring the points of CRINE to an onshore contractor", M.Sc. Thesis, Cranfield School of Management, Cranfield University.

Shaffer, L.R. and Micheau, J.W. (1971). "Bidding with Competitive Strategy Models", *Journal of the Construction Division*, Vol. 97, March. 113-126.

Shash, Ali (1998). "Bidding Practices of Subcontractors in Colorado" *J. Constr. Engrg. and Mgmt.* Vol. 124, May/June, 219-225.

Smith, F. (1997). "Bucking the trend: cost-plus services in lump sum turnkey market", *Journal of Management in Engineering*, 13, 38-43.

Songer, A.D. Molenaar, K.R. and Robinson, G.D. (1996). "Selection factors and success criteria for design-build in the U.S. and U.K." *J. Constr. Procurement*, Glamourgan, U.K, 2(2), 69-82.

Stark, R.M., (1968). "Unbalanced Bidding Models – Theory", *Journal of the Construction Division*, Vol. 94, October 1968, 197-209.

Szymanski, S., and Wilkins, S. (1993). "Cheap rubbish? Competitive tendering and contracting out in refuse collection-1981-88." *Fiscal Studies*, 14(3), 109-130.

Teicholz, P. and Ashley, D. (1978). "Optimal Bid Prices for Unit Price Contract" *Journal of the Construction Division*, Vol. 104, March, 57-67.

Tiong, R.L.K. (1998). "CSFs in Competitive Tendering and Negotiation Model for BOT projects" *J. Constr. Engrg. and Mgmt.* Vol. 124, Sept. 205-211.

Turner, D.F. (1975). "Building Contracts, A Practical Guide", George Godwin Limited, 2nd Edition.

United Nations (1973). "Guide on Drawing up Contracts for Large Industrial Work", (New York, United Nations, Economic Commission for Europe), 15-18.

U.K. Audit Office. (1987). "Competitive tendering for support services in the national health service (HC 318)." London: HMSO.

Veld, J. I. and Peeters, W.A. (1989). "Keeping large projects under control: the importance of contract type selection", *International Journal of Project Management*, 7, 155-162.

Ward, S. and Chapman, C. (1994). "Choosing contractor payment terms", *International Journal of Project Management*, 12, No. 4.

Yudha, S.W. (1995). "A cost reduction initiative for the oil and gas industry in Indonesia", M.Sc. Thesis, School of Industrial and Manufacturing Science, Cranfield University.

Chapter 5   Thou Shalt Not Be Ambiguous and Vague

Abraham, M.W. (1989). "Risk problems relating to construction". In *Construction Contract Policy—Improved Procedures and Practice*, eds. John Uff and Phillip Capper, Center of Construction Law and Management, King's College, London.

Abrahamson, M.W., (1983). "Risk Management". *Construction Insurance and Law*, FIDIC.

Abrahamson, M.W. (1979). "Engineering Law and the ICE Contracts", 4th ed. Applied Science, London, p. 218

Adsworth A. (1991). "Contractual Procedures in the Construction Industry", 2nd ed. Longman Scientific and Technical, London.

Barnes, N.M.L. (1996). "How contracts can help project managers", *Proceedings of the I.P.M.A. '96 Congress on Project Management*, June, Paris, France.

Barnes, Martin; Project Management/Deliotte and Sells Management Consultancy Division, (1987). "A New Style Contract for Engineering Contracts – A specification Prepared for the Institution of Civil Engineers".

Broome, J.C. and Hayes, R.W. (1997). "A comparison of the clarity of traditional construction contracts and of the New Engineering Contract" *International Journal of Project Management*, Vol. 15, No. 4, 255-261.

Caspe, M.S.; Igoe, I. and McDonald, S. (1991). "The Dispute Resolution Clause." *Project Management Journal*, March, 11-16.

Construction Industry Institute, CII. (1986). "Impact of Various Construction Contract Types and Clauses on Project Performance", Report# RS5-1.

Construction Industry Institute, CII. (1986). "Contract Clause Study Data", SD-9.

Construction Industry Institute, CII. (1986). "Determining the Impact of Various Construction Contract Types and Clauses on Project Performance", Vol. 1, Analysis and Recommendations, SD-10.

Construction Industry Institute, CII. (1986). "Determining the Impact of Various Construction Contract Types and Clauses on Project Performance", Vol. 11, Appendices, SD-11.

Construction Industry Institute, CII. (1986). "Analysis of Construction Contract Change Clauses", Vol.1, SD14, and Vol. 11, Sd-15.

Duncan Wallace, I. N. (1986). "Construction Contracts: Principles and Policies in Tort and Contract". Sweet and Maxwell, London, p.265.

Eggleston, B. (1993). "The ICE Conditions of Contract: A User's Guide", 6th ed. Blackwell Scientific Publications. Oxford

Green, L.J. (1982). "The Effect of Contract Wording on Inspector Liability", *ASCE Journal of Construction Engineering and Management*, March.

Hartman, F.T. (1994). "Reducing or Elimination Construction Claims by Changing Contracting Process", *Project Management Journal*, Vol. XXV No. 3, September, 25-31.

Hartman, F.T. (1998). "A tidal wave of weasel clauses", *Proceedings of the Annual Project Management Institute 1998 Seminars & Symposium*, Long Beach, CA, USA.

Hartman, F. and Snelgrove, P. (1996). "Effectiveness of Risk Allocation in Lump Sum Contracts-Concept of a Latent Dispute", *ASCE Journal of Construction Engineering and Management*, 122(3), Sept. 291-296.

Hartman, F., Snelgrove, P. and Ashrafi, R. (1997). " Effective Wording to Improve Risk Allocation in Lump Sum Contracts, *Journal of Construction Engineering and Management*, Vol. 123, No. 4, December, 379-387.

Jergeas, G.F. & Hartman, F.T. (1996). "A contract clause for allocating Risks", *AACE Transactions*, 1996, 1.1-1.3.

Latham, M., (1993). "Constructing the Team—Final Report of the Government Industry Review of Procurement and Contractual Arrangement in the UK Construction Industry" HMSO, London.

Oppenheim, A.N., (1992). "Question Design Interviewing & Attitude Measurement, Printer, London.

Rhys Jones, S. (1992). "The influence of law, language and perception on conflict in the construction industry." M.Sc. Dissertation, Center for Construction Law and Management, King's College, London.

Standard Construction Document CCDC2. (1982). "Stipulated price contract" Canadian Construction Documents Committee, Ottawa, Canada.

Standard Construction Document CCDC2 (1994). "Stipulated price contract" Canadian Construction Documents Committee, Ottawa, Canada.

Uff, J. (1991). Construction Law. Sweet & Maxwell, London, p. 216.

Valentine, D.G. (1992). "How not to draft a contract", *The International Construction Law Review*, 9, 526-529.

Chapter 6   Thou Shalt Share Out Risks Equitably and With Intelligence

Akintoye, A., and MacLeod, M. J. (1997). "Risk Analysis and Management in Construction", *International Journal of Project Management* 15, No. 1 1997, February: 31–38.

Al-Bahar, J.F., and Crandell, K. (1990). "Systematic Risk Management Approach for Construction Projects", *Journal of Construction and Management*, Vol. 116, No.3, September.

Aleshin, Artem. (2001). "Risk Management of International Projects in Russia", *International Journal of Project Management*, Vol.19, No.4, 207-222.

American Consulting Engineers Council & Associated General Contractors of America, Inc., (1990). "Owners Guide to Saving Money by Risk Allocation", Washington, DC.

Baccarini, D., and Archer, R. (2001). "The Risk Ranking of Projects: a methodology". *International Journal of Project Management*, Vol. 19, Issue 3, 139-145.

Baldry, D. (1998). "The evaluation of risk management in public sector capital projects", *International Journal of Project Management*, Vol. 16, No. 1, 35-41.

Berkley, D., Humphreys, P.C. and Thomas R.D. (1991). "Project Risk action management" *Construction Management and Economics*, 9, 3-17.

Bing, Li and Tiong, R.L.K. (1999). "Risk Management Model for International Construction Joint Ventures", *J. Constr. Engrg. and Mgmt*. 125, Sept/Oct., 377-384.

Bing, Li, Tiong, R.L.K, Wai Fan, and D. Ah-Seng. (1999). "Risk management in International Construction Joint Ventures", *J. Constr. Engrg. and Mgmt*. July/August, 277-284.

Birch, D.G.W. and McEvoy, M.A. (1992). "Risk analysis for information systems." *Journal of Information Technology*, 7:44-53.

Carter, B., Hancock, T., Morin, J, and Robin N. (1994). "Introducing RISKMAN: The European Project Risk Management Methodology" NCC Blackwell Limited, UK.

Casey, J.J. (1979). "Identification and Nature of Risks in Construction Projects: A Contractor's Perspective" *Proceedings of ASCE Construction Risks and Liability Sharing Conference*, Arizona, 17-23.

Chapman, C.B. (1992). "A risk engineering approach to risk management". In *Risk, Analysis, Assessment, and Management*, eds. J. Ansell and F, Wharton.Wiley, Chichester.

Chapman, Chris and Stephen Ward. (1997). "Project Risk Management", Chichester, John Wiley & Sons.

Chapman, R.J. (2001). "The controlling influences on effective risk identification and assessment for construction design management", *International Journal of Project Management*, Vol. 19, Issue 3, 147-160.

Chapman, R.J. (1998). 'The effectiveness of working group risk identification and assessment techniques", *International Journal of Project Management*, Vol. 16, No 6, 333-343.

Charoenngam, C. and Yeh, C. (1999). "Contractual Risk and Liability Sharing in Hydropower Construction', *International Journal of Project Management*, Vol. 17, No 1, 29-37.

Construction Industry Institute, CII. (1993). "Allocation of Insurance-Related Risks and Costs on Construction Projects", RS19-1.

Construction Industry Institute, CII. (1989). "Management of Project Risks and Uncertainties", RS6-8.

Construction Industry Institute, CII. (1994). "The Optimal Allocation of Insurance Related Risks and Costs in Construction Projects", SD-96.

Construction Industry Institute, CII. (1988). "Risk Management in Capital Projects".

Construction Industry Institute, CII. (1988). "Contract Risk Allocation and Cost Effectiveness", RS5-3.

Construction Industry Institute, CII. (1989). "Impact of Risk Allocation and equity in Construction Contracts", SD-44.

Conry, G.; and Soltan, H. (1998). "ConSERV, a project specific risk management concept", *International Journal of Project Management*, Vol. 16, No. 6, 353-366.

Dawson, RJ, CW Dawson, (1998). "Practical Proposals for managing uncertainty and risk in project planning". *International Journal of Project Management*, Vol. 16, No. 5, 299-310.

Del Cano. (1992). "Continuous Project Risk assessment" *International Journal of Project Management*, Vol. 10, p.165.

Dey, P., Tabucanon, T. and Ogunlana, S.O. (1994). "Planning for project control through risk analysis: a petroleum pipeline-layering project" *International Journal of Project Management*, 12:23-33.

Doherty, N. A. (1985). "Corporate Risk Management", McGraw-Hill Book Company, New York.

Edwards, Leslie. (1995). "Practical Risk Management in the Construction Industry." American Society of Civil Engineers (Thomas Telford, Ltd.).

Flanagan, R. and Norman, G., (1993). "Risk management and construction", Blackwell Scientific Publications, Oxford, UK.

Frantz, W. Forrest, (1990). "Systematic Risk Management Impacts Hybrid System Projects", Project Management Institute Seminar, October.

Gidfrey, P. "The Control of Risk, Risk, Management and Procurement in construction." Center of Construction Law and Management, King's College London.

Grey, S, (1995). "Practical Risk Assessment for Project Management", Wiley.

Hamburger, D. (1990). "The Project Manager: Risk Taker and Contingency Planner", *Project Management Journal*, June.

Harkunti, P.R., Carmichael, D.G. (1996). "A new Model of Risk Allocation for Construction Contracts based on Fair liabilities between parties. *ASCE Conference Proceedings, Computing in Civil Engineering*, 35-41.

Hartman, F. (1993). "Construction Dispute Reduction Through An Improved Contracting Process In the Canadian Context", Ph.D. Thesis, Loughborough University of Technology.

Hartman, F. (1994). "Reducing or Eliminating Construction Claims: A New Contracting Process (NCP). ", *Project Management Journal*, Vol. XXV, No. 3, September, 25-31.

Hartman, F.; Snelgrove, P. (1996). "Risk Allocation in Lump Sum Contract-Concept of a Latent Dispute", *ASCE Journal of Construction Engineering and Management*, Vol. 122, No. 3, Sept., 291-296.

Hartman, F., Snelgrove, P., Ashrafi, R. (1997). "Risk Allocation in Lump Sum Contact-Who should take which Risk?" *Cost Engineering*, July 1998. 21-26.

Hartman, F.T. Khan, Z., and Jergeas, G.F. (1997). "Understanding risk in construction contracts", *Proceedings of the PMI Annual Seminar & Symposium*, Chicago, 398-402.

Hartman, F.T. (1997). "Proactive Risk Management-Myth or Reality?" In *Managing Risks in Projects*, Edited by K. Kahnonen & K.A. Artto, E&FN Spon, England, 1997.

Head, G.L., and Horn II, S., (1991). "Elements of Risk Management", Insurance Institute of America.

Ireland, L. R., Shirley, V. D. (1986). "Measuring Risk in the Project Environment", Project Management Institute Seminar, September.

Isaac, A.G. (1990). "Analysis of Risks in Bidding Lump Sum Implementation Contracts", Project Management Institute Seminar, October.

Isaac, I. (1995). "Training in risk management", *International Journal of Project Management*, Vol. 13, No 4, 225-229.

Jafaari, Ali. (2001). "Management of risks, uncertainties and opportunities on projects: time for a fundamental shift". *International Journal of Project Management*, Vol. 19, Issue 2, 89-101

Jafaari, Ali, (1996). "Real time planning & Total Risk Management", *ASCE Conference Proceedings, Computing in Civil Engineering*, 193-199.

James, M. (Ed). (1996). "Risk Management in Civil, Mechanical and Structural Engineering", Thomas Telford.

Jergeas, G.F. & Hartman, (1995). "Method of Allocating Risk on Construction Contracts," in *Developments in Computer-Aided Design and Modeling for Civil Engineering*, Civil Comp Press, Cambridge, 1995, 69-74.

Kangari, R., and Boyer, L.T. (1986). "Risk Management by Expert Systems", *Project Management Journal*, March.

Kangari, R. (1995). "Risk Management Perceptions and trends of U.S. Construction" *J. Constr. Engrg. and Mgmt.* Vol. 121, Dec. 422-429.

Karlson, L. and Leiven, E., (1989). "Risk sharing Norwegian style." *Tunnel and Tunneling*, 33-35.

Khan, Zainul (1998). "Risk Premiums Associated with Exculpatory Clauses" Master of Science Thesis, University of Calgary Project Management Specialization.

Klien, J.H; and Cork, R.B. (1998). "An approach to technical risk assessment", *International Journal of Project Management*, Vol. 16, No.6, 1998, 345-351.

Kometa, S.T; Olomolaiye; P.O and Harris, F.C. (1996). "A review of client-generated risks to project consultants", *International Journal of Project Management*, Vol. 14, No.5, 1996, 273-279.

Lam, P.T.I. (1999). "A sectorial review of risks associated with major infrastructure projects." *International Journal of Project Management*, Vol. 17, No 2, 77-87.

Levitt, R. E. and Ashley, D. B. (1980). "Allocating Risk and Incentive in Construction", *Journal of Construction Engineering and Management*, September.

MacEwing, M. J. "Contractor's Assumption of Risk for Site Conditions", 21 CLR 257.

Macleod, M. J. (1994). "Perception and Management of Risk in the UK Construction Industry", Unpublished M.Sc. Dissertation, Glasgow, Caledonian University.

Martin, A.P., (1989). "Strategic Risk Analysis, Proactive Tools for Domestic and International Projects", Project Management Institute Seminar, October.

McKim, Robert A. (1992). "Risk Behavior of Contractors: A Canadian Study", *Project Management Journal*, Vol. XXII, No. 5, September, 51-55.

McKim R.A. (1992). "Systematic Risk Management Approach for Construction Projects." *J. of Constr. Engrg. & Mgmt.*, 118(2): 414-415.

McKim, R.A. (1992). "Risk Management-back to basics" *Cost Engineering*, 34(12), 712.

Mustafa, M.A. and Al-Bahar, J. (1991). "Project Risk Assessment Using the Analytical Hierarchy Process." *IEEE, Transactions on Engineering Management*, Vol. 38, No1, Feb. 46-52.

Noon, F. and Thain, H. (1986). "Identifying and assessing Risks for Projects", Project Management Institute Seminar, September.

Orman, G.A.E. (1991). "New applications of risk analysis in project insurances." *International Journal of Project Management*, Vol. 9, p. 131.

Orr, J. Barry. (1992). "Management of Risk - A Contractor's Viewpoint", Project Management Institute Seminar, September.

Peacock, W.S. and Whyte, E.I.L. (1982). "Site Investigation and Risk Analysis", *Proc. Civ. Engrs.*, May.

Pedwell, K; Hartman, F.T.; Jergeas, G.F. (1997). "Project capital cost risks and Contracting Strategies", *Journal of AACE International*.

Perry J.H. (1986). "Risk management-An approach for project managers." *International Journal of Project Management*, 211-216.

Porter, C.E. (1981). Risk allocation in construction contracts", M.Sc. thesis, University of Manchester.

Potter, M. (1995). "Procurement of Construction Work: The client's role." In Risk Management and Procurement in Construction, eds. J. Uff and A.M. Odams. *Proceedings of the 7th Annual Conference, Center of Construction Law and Management*. King's College, London, UK, 169-194.

Pym, D.V., and Wideman, R. Max, (1987). "Risk Management", *Project Management Journal*, August.

Raftery, J. (1994). "Risk Analysis in project management", E and F N Spon, London.

Raz, T; and Michael, E. (2001). "Use and benefits of project risk management", *International Journal of Project Management*, Vol. 19, Issue 1, 9-17.

Remenyi, Dan. (1999). "Stop IT project failures through risk management", Butterworth Heinemann.

Ren, H. (1994). "Risk Lifecycle and risk relationships on construction projects, *International Journal of Project Management*, Vol. 12, p.68.

Royal Aeronautical Society (1994). "Proceedings of the one day conference on Risk Management Techniques in Government and Industry". Royal Aeronautical Society, London.

Royal Society. (1992). "Risk: analysis, perception and management" Report to a Royal Society Study Group, London.

Simister, J. (1994). "Usage and benefits of project risk analysis and management", *International Journal of Project Management*, 12(1), 5-8.

Simon, P., Hillson, D. and Newland, K. (Eds). (1997). Project Risk Analysis and Management Guide, The Association of Project Managers, Norwich.

Smith, Gary, R and Bohn, Caryn. (2001). "Small to Medium Contractor Contingency and Assumption of Risk", *J. of Constr. Engg. and Mgmt*, 125, 76.

Snelgrove, P.N. (1994). "Risk Allocation in Lump-Sum Contracts". Master of Engineering Thesis, University of Calgary, Calgary, Alberta.

Stewrat, R.W. and Fortune, J. (1995). "Application of systems thinking to the identification, avoidance and prevention of risk" *International Journal of Project Management*, Vol. 13, No. 5, 279-286.

Strauss, M.W. (1979). Risks and Liability Sharing: The Owner's view. *Proceedings of ASCE Construction Risks and Liability Sharing Conference*, Arizona, 25-33.

Tao, C.W. (1994). "Needs for equitable risk sharing contracts" Sonotech Engineering, 44, 3-11.

Thompson, P.A., and Perry, J.G. (1992). "Engineering Construction Risk", Thomas Telford Services Ltd., London.

Titarenko, B.P. (1997). "Robust technology in risk management", *International Journal of Project Management*, Vol. 15, No.1, 11-14.

Toakley. (1995). "Risk Management applications- a review" Australian Inst. of Build. Papers 6, 77-85.

Tummala, V.M.R., Nkasu, M.M. and Chuah, K.B. (1994). "A Systematic Approach to risk management" *Journal of Mathematical Modeling and Scientific Computing*, No. 4, p.174-184.

Tummala, V.M.R., Nkasu, M.M. and Chuah, K.B. (1994). "A framework for project risk management." *ME Research Bulletin*. No.2, 145-171.

Tummala, V.M.R. and Burchett, J.F. (1999). "Applying a risk management process (RMP). to manage cost risk for an EHV transmission line project", *International Journal of Project Management*, Vol. 17, No. 4, 223-235.

Uff, J. (1989). "Origins and development of construction contracts". In *Construction Contract Policy—Improved Procedure and Practice*, eds. John Uff and Phillip Capper, Center of Construction Law and Management, King's College, London.

Uher, T.E.; AR Toakley. (1999). "Risk Management in the Conceptual Phase of a Project", *International Journal of Project Management*, Vol. 17, No. 3, 161-169.

Ward, SC. (1999). "Assessing and managing important risks", *International Journal of Project Management*, Vol. 17, No. 6, 331-336.

Ward, S. C. and Chapman, C. B. (1991). "Extending the use of Risk Analysis In Project Management", *Project Management Journal*, Vol.9 No.2, May.

Ward, S.C, Chapman, C B, and Curtis, B. (1991). "On the allocation of risk in construction projects" *International Journal of Project Management*, Vol. 9, No. 3, p.140-147.

Ward, S.C; C.B. Chapman. (1995). "Risk-management perspective on the project lifecycle", *International Journal of Project Management*, Vol. 13, No. 3, 145-149.

Ward, Stephen, (1999). "Requirements for an effective Project Risk Management Process", *Project Management Journal*, September, 37-43.

Wharton, F. (1992). "Risk management: basic concepts and general principles." In *Risk, Analysis, Assessment and Management*, eds. J. Ansell and F. Wharton. Wiley, Chichester.

William, T.M. (1993). "Risk-Management infrastructures" *International Journal of Project Management*, Vol. 11, p. 68.

William, T.M. (1994). "Using a risk register to integrate risk management in project definition. *International Journal of Project Management*, Vol. 12, p. 17-24

Williams, T.M. (1997). "Empowerment vs. risk management?" *International Journal of Project Management*, Vol. 15, No. 4, 219-222.

Williams, T.M. (1996). "The two-dimensionality of project risk", *International Journal of Project Management*, Vol.14, No. 3:185-186.

Yeh, C.Y. (1995). "Construction Risks and liability sharing in Taiwanese hydropower project contract" Master Thesis, Asian Institute of Technology. Thailand.

Yeo, K.T. (1990). "Risks, Classifications of Estimates, and Contingency Management", *Journal of Management in Engineering,* Vol.6, October.

Yeo, K.T. (1990). "Risk classification of estimates and contingency" *Journal of Engineering Management*, 6, 458-470.

Yeo, K.T., Tiong, R.L.K. (2000). "Positive management of differences for risk reduction in BOT projects", *International Journal of Project Management*, Vol. 18, No.4, 257-265.

Yosua, D. A. and Hazlett, R. L. (1988). "Risk Management, The Proposed Standard for Department of Defense Program Managers", Project Management Institute Seminar, September.

YS, L. (1997). Project risk management in Hong Kong" *International Journal of Project Management*, Vol. 15, No. 2, 101-105.

Zhi, H. (1995). "Risk Management for Overseas construction projects", *International Journal of Project Management*, Vol. 13, No. 4, 231-237.

Chapter 7   Thou Shalt Trust thy Contracting Partner, but not do so Unreasonably

Baier, Anette. (1985). "Trust and Antitrust", *Ethics*, 96: 231-260.

Barber, B. (1983). "The logic and limits of Trust". New Brunswick, NJ: Rutgers University Press.

Barney, J.B, and M.H. Hansen. (1994). "Trustworthiness as a Source of Competitive Advantage." *Strategic Management Journal* 15:175-190.

Bies, Robert. J. and Tripp, Thomas .M. (1996). "Beyond Distrust in Trust in Organizations." In Roderick M. Kramer and T.R. Tyler (Eds.). *Trust in organizations: Frontiers of theory and research*. Thousand Oaks, CA: Sage Publications: 246-260.

Bradach, J.L. and Eccles, R.G. (1989). "Price Authority and Trust. From ideal types to plural forms", *Annual Review of Sociology*, 15, 97-118.

Brien, Andrew. (1998). "Professional ethics and the culture of trust", *Journal of Business Ethics*, Vol. 17(4): 391-409.

Brenkert, George G. (1998). "Trust, morality and international business", *Business Ethics Quarterly*, Vol. 8(2): 293-317.

Butler, John K. Jr. (1991). "Towards understanding and measuring conditions of trust: Evolution of a trust condition inventory". *Journal of Management*, Vol. 17: 643-663.

Carnevale, D.G. (1995). Trustworthy Government: Leadership and Management Strategies for Building Trust and High Performance.", San Francisco: Jossey-Bass.

Construction Industry Institute, CII. (1993). "Cost-Trust Relationship", Publication No. 24-1, CII Contracting Phase II Taskforce.

Construction Industry Institute, CII. (1994). "The Cost-Trust Relationship in the Construction Industry", SD100.

Couch, Laurie L. and Jones Warren H. (1997). "Measuring Levels of Trust", *Journal of Research in Personality*, Vol. 31: 319-336.

Coutu, Diane L. (1998). "Organizations: Trust in virtual teams". *Harvard Business Review*, Vol. 76(3): 20-21.

Creed, W.E. Douglas and Miles, Raymond E. (1996). "Trust in Organizations: A conceptual framework linking organizational forms, managerial philosophy, and the opportunity costs of controls." In Roderick M. Kramer and T.R. Tyler (Eds.). *Trust in organizations: Frontiers of theory and research*. Thousand Oaks, CA: Sage Publications: 16-38.

Farries G., Senner, E. and Butterfield, D. (1973). "Trust, Culture and organizational behavior", *Industrial Relations*, 12: 144-157

Flores Fernando and Solomon Robert C. (1998). "Creating Trust". *Business Ethics*, Vol. 8(2), 205-232.

Fukuyama F., (1995). "Trust: The Social Virtues and the creation of prosperity". New York: McMillan.

Fukuyama, F. (1995). "Trust." Free Press, New York.

Gabarro, J.J. (1978). "The development of trust influence and expectations." In A.G. Athos & J.J. Gabarro (Eds.), *Interpersonal behavior: Communication and understanding in relationships*, Englewood Cliffs, N.J.: Prentice-Hall, 290-303.

Gambetta, D. (1988). "Can We Trust", in D. Gambetta (ed.) *Trust: Making and Breaking Cooperative Relationships*." Basil Blackwell, New York.

Gulati, R. (1995). "Does Familiarity Breed Trust? The Implications of Repeated Ties for Contractual Choice in Alliances." *Academy of Management Journal,* 38:85-112.

Hartman, F.T. (1999). "The Role of Trust in Project Management, *Proceedings of Nordnet '99, International Project Management Conference*, Helsinki, Finland, September.

Hartman, F.T. (1999). "Report on the Development of Project Management Research into Trust", *Proceedings of Project Management Vienna VI Conference*, Vienna, Nov.

Hartman, F.T. (2000). "Research into Trust in Project Management", *Proceedings of the Vienna Project Management Research Network Conference*, Vienna, Austria, November.

Hartman, F.T. (2000). "The Role of Trust in Project Management", *Proceedings of the PMI Research Conference*, Paris, France, June.

Hartman. F.T. (2000). "Trust-based Contracts – a competitive advantage or a high-risk proposition?" *Proceedings of the European Project Management Conference*, Jerusalem, Israel, June.

Hartman, F.T. (2000). "Smart Trust – a foundation for more effective project management application", *Proceedings of the 15th IPMA World Congress on Project Management (Congress 2000)*, May, London, United Kingdom.

Hartman, F.T. and Romahn, E. (2000). "The Role of Trust in Project Success", *Proceedings of IRNOP IV (International Research Network on Organizing by Projects) Conference*, Sydney, Australia, January.

Held V. (1984). "Rights and goods: Justifying social action." New York and London: Free press.

Herzog, V. (2000). "Building Collaborative Trust", M.Sc. Thesis, Project Management, Civil Engineering, University of Calgary.

Johnson-George, C. and Swap, W.C. (1982). "Measurement of specific interpersonal trust: construction and validation of a scale to assess trust in a specific other" *J Personality & Social Psychology*, Vol. 43, 1306-1317.

Kipnis, David. (1996). "Trust and Technology". In Roderick M. Kramer and T.R. Tyler (Eds.). *Trust in organizations: Frontiers of theory and research*. Thousand Oaks, CA: Sage Publications: 39-49

Koehn, Daryl. (1996). "Should we trust in trust." *American Business Law Journal*, Vol. 34: 183-203.

Kramer, R.M. and T.R. Tyler (eds.). (1996). *"Trust in organizations: Frontiers of theory and research."* Thousand Oaks, CA: Sage.

LaPorta, R., Lopez, F., Shleifer, A. and Vishni, R. (1997). "Trust in Large Organizations." AEA Papers Proceedings.

Latham, Sir M. (1993). "Trust and Money. Interim report of the Government/Industry review of procurement and contractual arrangements in the UK construction industry." HMSO, London.

Lewicki, Roy J. and Benedict Bunker, Barbara. (1996). "Developing and maintaining trust in work relationships", In Roderick M. Kramer and T.R. Tyler (Eds.). *Trust in organizations: Frontiers of theory and research*. Thousand Oaks, CA: Sage Publications: 114-139.

Lindskold, S. (1978). "Trust development, the GRIT proposal, and the effects of conciliatory acts on conflict and cooperation" *Psychological Bulletin*, 85(4), 772-793.

Mayer, R.C., Davis, J.H. and Schoorman F.D. (1996). "An Integrative Model of Organizational Trust." *Academy of Management Review*, 20:709-734.

McAllister, D.J. (1995). "Affect-and Cognition-Based Trust as Foundations for Interpersonal Cooperation in Organizations." *Academy of Management Journal*, 38:24-59.

McKnight, D. Harrison, Cummings, Larry L. and Chervany, Norman L. (1998). "Initial trust formation in new organizational relationships." *Academy of Management Review*, Vol. 23(3): 473-490.

Meyerson, Debra, Weick, Karl E. and Kramer, Roderick M. (1996). "Swift Trust and Temporary Groups." In Roderick M. Kramer and T.R. Tyler (Eds.). *Trust in organizations: Frontiers of theory and research*. Thousand Oaks, CA: Sage Publications: 166-195.

Mishra, Aneil. (1996). "Organizational Responses to Crisis: The Centrality of Trust", In Roderick M. Kramer and T.R. Tyler (Eds.). *Trust in organizations: Frontiers of theory and research*. Thousand Oaks, CA: Sage Publications: (140-165).

Mishra, Aneil K. and Spreitzer, Gretchen M. (1998). "Explaining how survivors respond to downsizing: The role of trust, empowerment, justice, and work redesign." *Academy of Management Review*, Vol. 23(3): 567-588.

Munns, A.K. (1995). "Potential influence of trust on the successful completion of a project", *International Journal of Project Management*, Vol. 13, No 1, 1995, 19-24.

Romahn, E; Hartman, F. (1999). "Trust, a New Tool for Project Managers", *Proceedings of the Project Management Institute 1999 Annual Symposium and Seminar*, Philadelphia, Pennsylvania. October.

Rotter, Julian .B. (1967). "A new scale for the measurement of interpersonal trust", *Journal of Personality*, Vol. 35: 615-665.

Sako, M. (1991). "Role of Trust in Japanese Buyer-Supplier Relationships", *Ricerche Economiche*, 45/2-3 (April-September): 449-474.

Sheppard, Blair H. and Sherman, Dana M. (1998). "The grammar of trust: A model and general implications." *Academy of Management Review*, Vol. 23(3): 422-437.

Sitkin, S.N. and Roth, N.L. (1993). "Explaining the Limited Effectiveness of Legalistic 'Remedies' for Trust/Distrust." *Organization Science* 4:367-392.

Zand, D.E. (1972). "Trust and managerial problem solving" *Administrative Science Quarterly*, 17, 229-239.

Zucker, L.G. (1986). "Production of trust: Institutional sources of economic structure." 1840-1920, in B.W. Staw and L.L. Cummings (Eds.), *Research in Organizational Behavior*, Vol. 8: 53-111.

Zucker, A., McEvily, B. and Perrone, V. (1996). "Does Trust Matter? Exploring the International and Interpersonal Trust on Performance." Working Paper, University of Minnesota, Minneapolis.

Chapter 8   Thou Shalt Not Mess Unduly with the Contract
After it is Agreed

Appleton. (1987). "Town of Slave Lake v. Appleton Construction Ltd. and Western Surety Company." 25 C.L.R. 311 (Alta. Q.B.)

ASCE and Santiago, S.J. (1997). "Patent Verses latent Contract Ambiguities: Who Bears the Increased Costs of Performance?" *Civil Engineering-ASCE*, Vol. 67, No. 9, September, p. 69.

Bristow, D.I. (1996). "Damages Arising from the Bidding Process." Unpublished manuscript.

Brown, D. (1990). "Contract drafting: a risky business" *Consulting Specifying Eng.* 29-30.

Bubshait, A.A. and Almohawis, S.A. (1994). "Evaluating the general conditions of a construction contract" *International Journal of Project Management*, Vol. 12, 133.

Clough, R. (1986). "Construction Contracting" (5th Ed.). John Wiley, USA

Currie, O.A. and Dorris, W.E. (1986). "Understanding construction contracts: a myriad of special clauses" *Arbitration J.,* Vol. 41 No 1, 3-16.

Fisk, E.R. (1998). "Construction Project Administration" (2nd Ed.). John Wiley, USA.

Gilberth, R. (1983) "Managing Construction Contracts" John Wiley, USA.

Goodfellow, W.D. (1995). "Law of Tendering." Unpublished manuscript.

Goudsmit, J.J. (1990). "Exclusion or Limitation of Liability in a Turnkey Contract." *The International Construction Law Review*, Vol. 7, Part 3, July. Lloyd's of London Press Ltd.

Groton, J. (1986). "A guide to improving construction contracts" *Arbitration J.* Vol. 41 No 2, 21-24.

Ibbs, W. and Ashely, D. (1987). "Impact of various construction contract clauses." *J. Construct. & Manage.* Vol. 113, No 3, 501-521.

Jergeas, G.F. and Cooke, V.G. (1997). "Law of Tender Applied to Request for Proposal Process", *Project Management Journal*, Dec. Vol.28, No 4.

Jervis, B.M. and Levin, P. (1988). "Construction Law: principles and Practice", McGraw-Hill, N.Y.

Jack, H. (1996). "The Professional Practice of Engineering."

Kangari, R. (1995). "Construction Documentation in Arbitration." *J. Constr. Engineering and Management*. Vol. 121, June, 201-208.

Loulakis, M.C. and Cregger, W.L. (1996). "The Importance of Contract Clarity Clarified." *Civil Engineering*, ASCE, Vol. 66, No.3, March, p.32.

Marston, D. L. (1985). "Law for Professional Engineers." Second Edition. Toronto: McGraw-Hill Ryserson.

Ndekugri, I. (1999). "Performance Bonds and Guarantees: Construction Owners and Professions Beware." *J. Constr. Engrg. and Mgmt.* 125, Nov/Dec., 428-436.

Northern. (1984). "Northern Construction Ltd. v. The City of Calgary." 52 A.R. 54, (1986). 67 A.R. (Alta C.A.). 150

Ron. (1981). "Her Majesty the Queen in Right of Ontario v. Ron Engineering and Construction (Eastern). Ltd." 119 D.L.R (3d). 267 (S.C.C.)

Roth, L. A. (1992). "Legal Corner: Questions and Answers on Construction Law in Ontario." Cassels, Brock and Blackwell Newsletter, June.

Roth, L. A. (1995). "Construction and the Law." Cassels, Brock and Blackwell Newsletter, Spring.

Schoumacher, Bruce, H. (1997). "Contract Awards and Contract Administration, Conference proceedings, Quality assurance-A national Commitment, 216-224.

Semple, C. Hartman, F.T., Jergeas, J. (1994). "Construction claims and disputes: cause and cost/time overruns," *ASCE Journal of Construction Engineering and Management*, 120(4); 785-795.

Stuart-Ranchev, A. (1990). "Design Responsibility." *Construction Dispute-Avoidance and Resolution Conference Proceedings*, Law and Business Forum, September.

Sweet, J. (1989). "The use of standard forms for construction management contracts" *Real Estate Law J.* Vol. 18, No 1, 74-89.

Thomas, H.R. and Smith, G.R. and Wright, D.E. (1991). "Legal aspects of Oral Change Orders." *Journal of Construction Engineering and Management*, Vol. 117, March, 148-162.

Thomas, H.R, Smith G.R., and Ponderlick, R.M. (1992). "Resolving Contract Disputes Based on Differing-Site condition Clause." *Journal of Construction Engineering and Management*, Vol. 118, Dec., 767-779.

Thomas, H.R, Smith G.R., and Ponderlick, R.M. (1992). "Resolving Contract Disputes Based on Misrepresentations." *Journal of Construction Engineering and Management*, Vol. 118, Sept., 472-487.

Thomas, H.R., Smith, G.R., Mellott, R.E. (1994). "Interpretation of Construction Contracts", *Journal of Construction Engineering and Management*, Vol. 120, No.2, June, 321-336.

Thompson, P. A. and Perry, J.G. (Eds.). (1992). "Engineering Construction Risks – A Guide of project Risk Analysis and Risk Management" Thomas Telford, UK.

Uff, J. (1991). "The ICE Conditions of Contract" 6th Ed. New thoughts on Old Issues. *The International Construction Law Review*, Vol. 8, part 2, April, Lloyds of London Press Ltd.

Wearne, S.H. (1992). "Contract Administration and project risks", *International Journal of Project Management*, Vol. 10, p. 39.

Chapter 9   Thou Shalt Deal Rationally and Fairly with Inevitable Changes to a Contract

Allen, W.E. (1993). "A methodology fro evaluating the effective processing of project change." Dissertation Proposal, University of California at Berkeley Construction Management program.

Al-Sedairy, Salman, T. (2001). "A change management model for Saudi Construction industry." *International Journal of Project Management*, Vol. 19, No.3, 161-169.

Al-Tabtabai, Hashem, M. (1999). "Change orders in Construction Project: Causes, effects and Solutions", *Project Management Institute 30th Annual Seminars and Symposium Proceedings*, Oct., Philadelphia, PA, USA.

Assaf, S.A.; Bubshait, A.A.; Atiyah, S. and Al-Shahri, M. (2001). "The management of construction company overhead costs." *International Journal of Project Management*, Vol.19, Issue 5, July, 295-303.

Boddy, David; Macbeth, Douglas. (2000). "Prescriptions for managing change: a survey of their effects in projects to implement collaborative working between organizations." *International Journal of Project Management*, Vol. 18, Issue 5, 297-306.

Bubshait, A. and Cunningham, M.J. (1998). "Comparison of Delay Analysis Methodologies" *J. Constr. Engrg. and Mgmt.* Vol. 124, July/Aug., 315-322.

Christensen, D.S. (1993). "An Analysis of Cost overruns on Defense Acquisition Contracts", *Project Management Journal*, September, p. 43-48.

Coe, C. K., & O'Sullivan, E. (1993). "Accounting for the hidden costs; A national study of internal service funds and other indirect costing methods in municipal governments." *Public Administration Review*, 53, 59-64.

Construction Industry Institute (CII). (1994). "Project Change Management." SP43-1.

Construction Industry Institute (CII). (2000). "Quantifying the Cumulative Impact of Change Orders for Electrical and Mechanical Contractors." RS158-1.

Construction Industry Institute (CII). (1994). "Quantitative Effects of Project Change", RS43-2.

Construction Industry Institute (CII). (1990). "The Impacts of Changes on Construction Cost and Schedule", RS6-10.

Construction Industry Institute (CII). (1995). "Quantitative Effects of Project Change" SD-108.

Construction Industry Institute (CII). (1991). "Construction Changes and Change Orders: Their Magnitude and Impact." SD-66.

Construction Industry Institute (CII). "Early Warning Signs of Project Changes" SD-91.

Diekmann, J.E. and Kim, M.P. (1992). "SuperChange: Expert System for Analysis of Changes Claims." *Journal of Construction Engineering and Management*, Vol. 118, June, p. 399-411.

Dozzi, S.P.; AbouRizk, S.M. and Schroeder, S.L. (1996). "Utility-Theory Model for Bid Markup Decisions" *J. Constr. Engrg. and Mgmt*. Vol. 122, June, 119-124.

Dye, R. A. (1985). "Costly Contract Contingencies." *International Economic Review*, 26, 233-250.

Ehrenreich-Hansen, F. (1994). "Change order management for construction projects", *Cost Engineering*, 36(3), 25-28.

Green, J., and Laffont, J.J. (1992). "Renegotiation and the Form of Efficient Contracts." *Annales d'Economie et de Statistique*, 25/26:123-150.

Griffis, F.H. and Christodoulou, S. (2000). "Construction Risk Analysis tool for determining Liquidated damages Insurance Premiums: Case Study." *J. Constr. Engrg. and Mgmt*. 126, Nov/Dec, 407-413.

Huntoon, Carolyn, L. (1998). "Managing Change", *Project Management Journal*, September, 5-6.

Ibbs, C. W. and Allen, W.E. (1995). "Quantitative impacts of project change" University of California at Berkeley, Construction Management Technical Report No.23.

Ibbs, C.W, (1997). "Change's impact on construction productivity" *ASCE Journal of Construction Engineering and Management*, 123(1), 89-97.

Ibbs, C.W. (1998). "Quantitative Impacts of Project Change: Size issues" *J. Constr. Engrg. and Mgmt*. 123, Sept., 308-311.

Ibbs, C. W., Lee, S.A. and Li, M. I. (1998). "Fast Tracking's impact on Project Change" *Project Management Journal*, December, 35-41.

Kallo, G.G. (1996). "The reliability of critical path method (CPM). techniques in the analysis and evaluation of delay claims." *Cost Engineering*, 38(5), 35-37.

Krone, S.J. (1992). "Modeling Construction Change Order Management", *Project Management Journal*, September, 17-19, 35.

Li, H., Shen, Y. and Love, P.E.D. (1999). "ANN-based Mark-Up Estimation System with Self-Explanatory Capacities." *J. Constr. Engrg. and Mgmt*. 125, May/June, p.185-189.

Mak, Stephen and Picken, D. (2000). "Using Risk Analysis to determine Construction Project Contingencies." *J. Constr. Engrg. and Mgmt*. 126, Mar/Apr. p. 130-136.

Maskin, E., and Tirole, J. (1999). "Unforeseen Contingencies and Incomplete Contracts." *Review of Economic Studies*, 66, 83-114.

Newman, Robert B., and Frederick D. Hejl. (1984). "Contracting Practices and Payment Procedures." *Transportation Research Record,* 986: 50-59.

Salerno, V. (1995). "Managing project change", Project Management Institute's 26th Annual Seminar and Symposium. Chicago. Project Management Institute.

Smith, G.R. and Bohn, C.M. (1999). "Small to medium Contractor Contingency and Assumption of Risk" *J. Constr. Engrg. and Mgmt*. 125, Mar/Apr. p.101-108.

Tavakoli, A. and Utomo, J.L. (1989). "Bid markup assistant: An expert system." *Cost Engineering*. 31(6), 28-33.

Voropajev. V. (1998). "Change Management—A key integrative function of PM in Transition economies." *International Journal of Project Management*, Vol. 16, No. 1, 15-19.

William Ibbs, C; Lee, Stephanie and Li, Michael I. (1998). "Fast-Tracking's impact on Project Change." *Project Management Journal*, December, 35-41.

442

Chapter 10 Thou Shalt Not Kill (Your Contracting Partner)

Abourizk, S.M and Dozzi, S.P. (1993). "Application of Computer Simulation in Resolving Construction Disputes." *Journal of Construction Engineering and Management*, Vol. 119, June, p. 355-373.

Abudayyeh, O.Y. (1997). "A multimedia construction delay management system" Microcomp. *Civil Engineering*. 12, 183-192.

Alkass, S., Mazerolle, M., Tribaldos, E. and Harris, F. (1995). "Computer aided construction delay analysis and claims preparation." *Constr. Mgmt. and Economics*, 13, 335-352.

Al-Khalil, M.I., and Al-Ghafly, M.A. (1999), "Delay in public utility projects in Saudi Arabia. *International Journal of Project Management*, Vol. 17, No. 2, 101-106.

Al-Momani, Ayman, H. (2000). "Construction delay: a Quantitative analysis" *International Journal of Project Management*, Vol.18, No.1, 51-5 9.

Al-Sedairy, S T (1994). "Management of Conflict. Public Sector construction in Saudi Arabia" *International Journal of Project Management*, Vol. 12, p.143.

American Society of Civil Engineers (1991). "Avoiding and Resolving Disputes During Construction: Successful Practices and Guidelines", ASCE.

Arditi, D. and Robinson, M.A. (1995). "Concurrent delays in Construction litigation." *Cost Engineering*. 37(7), 20-30.

Assaf, S.A., Al-Khalil, M., and Al-Hazmi, M. (1995). "Causes of delay in large building construction projects" *J. Management in Engineering*, ASCE, 11(2), 45-50.

Bartsch, S. and Jergeas, G. (2000). "An Exploration of Claims in the Construction Industry." *Proceedings of CSCE 2000 Conference*, Calgary, June.

Bartsch, S. Jergeas, G. and Hartman, F. (2000). "Avoiding Claims and Disputes in the Construction Industry" *Proceedings of the European Project Management Conference*, Jerusalem, Israel, June.

Battikha, M, and Alkass, S. (1994). "Cost-effective delay analysis technique." *Transactions*, American Association of Cost Engineers (now AACEI). Morgantown, W.Va.

Carroll, E. and Dixion, G. (1990). "Alternate Dispute Resolution Development in London" *The International Construction Law Review*, 7(4), 436-442.

Casner, H.D. (1988). "The Cost Engineer and Claims Prevention." AACE International's Professional Practice Guide to Contracts and Claims, *AACEI Transactions*.

Cheung, S. (1999). "Critical factors affecting the use of alternate dispute resolution process in construction." *International Journal of Project Management*, Vol. 17, No.3, 189-194.

Cheung, S. and Yeung, Y. (1998). "The effectiveness of the Dispute Resolution Advisor System: a critical appraisal." *International Journal of Project Management*, Vol. 16, No. 6, 367-374.

Clark, William G. (1990). "Claims Avoidance and Resolution" AACE International's Professional Practice Guide to Contracts and Claims, *AACEI Transactions*.

Collins, A. (1988). "Development in Alternate Dispute resolution in North America." *Arbitration*, 54(4), 242-251 and 262.

Conlin, J. and Retik, A. (1997). "The applicability of project management software and advanced IT techniques in construction delays mitigation" *International Journal of Project Management*, Vol. 15, No. 2, 107-120.

Construction Industry Institute, CII. (1994). "DPI-Dispute Potential Index: A Study into the Predictability of Contract Disputes." SD-101.

Construction Industry Institute, CII. (1993). "Dispute Prevention and Resolution" SD-95.

CPR Legal Program Construction Disputes Committee. (1991). "Preventing and Resolving Construction Disputes." Center for Public Resources, Inc.

DiDonato, S.L. (1993). "Contract Disputes: Part 1, Alternatives for Dispute Resolution", *PM Network*, May, 19-23.

DiDonato, S.L. (1993). "Contract Disputes: Part II, Dispute Avoidance." *PM Network*, August, 53-57.

Diekman, J.E. and Girard, M. (1995). "Construction Industry Attitudes towards Disputes and Prevention/resolution Techniques", *Project Management Journal*, March. 3-11.

Dikemann, J.E. and Girard, M. J. (1995). "Are Contract Disputes Predictable? *J. Construction Engineering and Management*. Vol. 121, Dec., 355-363.

Fredlund, Donald J. and others (1989). "Innovative Contract Documents and Progressive Project Management Yields Success." AACE International's Professional Practice Guide to Contracts and Claims, *AACE Transactions*.

Goldberg, S.B., Sander, F.E.A. and Rogers, N.H. (1992). "Dispute Resolution", 2nd Edn. Little Brown and Company, USA.

Groton, J.P. (1992). "Supplementary to Alternative Dispute Resolution in the Construction Industry" Wiley Law Publications, USA.

Hanbury, W. (1992). "Alternative dispute resolution-the Australian model." *Solicitors Journal*, April, 334-335.

Hartman, F.T. (1993). "Construction Dispute Resolution Through an Improved Contracting Process in the Canadian Context", Ph.D. Thesis, Loughborough University of Technology, Loughborough, UK.

Hartman, F.T. (1994). "Reducing or Eliminating Construction Claims: a New Contracting Process (NCP)". *Project Management Journal*, Vol. XXV, No. 3, 25-31.

Hartman, F.T and Jergeas G.F. (1995). "A model for proactive mediation of construction Disputes", *Canadian Journal of Civil Engineering*, 22(1); 14-22.

Henderson, T.R. (1991). "Analysis of Construction Disputes and Strategies for Cost Minimization" *Cost Engineering*, Vol.33, No11, 17-21.

Jannadia, M. Osama; Saadi Assaf; A.A. Bubshait; Alam Naji. (2000). "Contractual methods for dispute avoidance and resolution (DAR)", *International Journal of Project Management*, Vol.18, No. 1; 2000, 41-49.

Jensen, D. A., Murphy, J.D., and Craig, J. Jr. (1997). "The Seven Legal Elements Necessary for a Successful Claim for a Constructive Acceleration." *Project Management Journal* 28, No. 1 (March): 32-44.

Jergeas, G. F. and Hartman, F.T. (1994). "Contractor's protection against Construction Claims", *AACEI Transactions*, 8.1-8.5.

Jergeas, G.F. and Hartman, F.T. (1994). "Contractors construction claims avoidance." ASCE *Journal of Construction Engineering and Management*, 120(3); 553-560.

Jergeas, G.F. Revay, S.O. (1993). "Quantifying Construction Claims Using the Differential Method" *Construction Management & Economics*, E&FN Spon, UK, 163 - 166.

Jergeas, G.F. and Hartman, F.T. (1996). "A Contract Clause for Allocating Risk." AACE International's Professional Practice Guide to Contracts and Claims, *AACEI Transactions*.

Kallo, Gasan. (1990). "Claims Management." *Cost Engineering* 32, No.10 (October): 25–26.

Kartam, S. (1999). "Generic Methodology for Analyzing Delay Claims." *J. Construction Engineering and Management*. 125, Nov./Dec. 409-419.

Leishman, D.M. (1991). "Protecting Engineer Against Construction Delay claims: NDC." *Journal of Management in Engineering*, Vol. 7, No.3, July, 314-333.

Levin, Paul (1998). "Construction Contract Claims, Changes & Dispute Resolution" 2nd Edition, American Society of Civil Engineers (ASCE Press).

Lim, C.S; M. Zain Mohamed. (2000). "An exploratory study into recurring construction problems." *International Journal of Project Management*, Vol. 18, No 4, 2000, 267-273.

Mansfield, N.R., Ugwu, O and Doran, T. (1994). "Causes of delay and cost overruns in Nigeria construction projects" *International Journal of Project Management*, Vol. 12, 254.

Matyas, R.M., P.E. Sperry, R.J. Smith, and A.A. Mathews. (1996). "Construction Dispute Review Board", McGraw Hill.

Molenaar, Keith; Washington, Simon and James Diekmann, (2000). "Structural Equation Model of Construction Contract Dispute Potential" *J. Constr. Engrg. and Mgmt.* 126, 268-277.

National Public Works Conference and National Building and Construction Council Report. (1990). "No Dispute: Strategies for Improvements in the Australian Building and Construction Industry." Pirie Printers Sales. Australia.

Naughton, P. (1990). "Alternate forms of dispute resolution-their strengths and weaknesses" *Construction Law Journal*, 6(3), 195-206.

Negotiation and Contract Management. (1985) *Proceedings of the Symposium Sponsored by the Engineering Management Division of the American Society of Civil Engineers in Conjunction with the ASCE Convention*, Denver, Colorado, April 29-30.

Ogunlana, S.O; Promkuntong, K. and Jearkjirm, V. (1996). "Construction delays in a fast-growing economy: comparing Thailand with other economies." *International Journal of Project Management*, Vol. 14, No 1, 37-45.

Pears, G. (1989). "Beyond Dispute-Alternative Dispute Resolution in Australia" Corporate Impacts Publications, Australia.

Pinto, J.K., Kharbanda. O.P. (1995). "Project Management and Conflict Resolution." *Project Management Journal*, March, 21-31.

Poulin, Thomas A. Editor. (1985). "Avoiding Contract Disputes."
     *Proceedings of a Symposium sponsored by the Construction Division
     of the American Society of Civil Engineers in conjunction with the
     ASCE Convention*, Detroit, Mich., Oct. 21-22, American Society of
     Civil Engineers.

Pressoir, Serge (1992). "Are You Contracting for Success?" AACE
     International's Professional Practice Guide to Contracts and Claims,
     AACE Transactions.

Quick, R.W. (1993). "How to Avoid Litigation", *PM Network*, February,
     33-36.

Revay, S.O. (1994). "On Site Dispute Resolution" *The Revay Report*,
     16:4.

Russell, Jeffrey S, (2000). "Surety Bonds for Construction Contracts."
     American Society of Civil Engineers (ASCE Press).

Sandori, P. (1994). "Avoiding the Courtroom battle of Experts", *PM
     Network*, February, 42-43.

Schumacher, L. (1995). "Quantifying and apportioning delay on
     construction project" *Cost Engineering*, 37(2), 11-13.

Scott, S. (1993). "Dealing with delay claims: a survey" *International
     Journal of Project Management*, Vol. 11, p. 143.

Scott, S. (1997). "Delay Claims in UK Contracts" *J. Constr. Engrg. and
     Mgmt*. Vol. 123, Sept., 238-244.

Semple, C., Hartman, F.T. and Jergeas, G. (1994). "Construction claims
     and disputes: cause and cost/time overruns," *ASCE Journal of
     Construction Engineering and Management*, 120(4); 785-795.

Severson, G.D.; Russell, J.S. and Jaselskis, E.J. (1994). "Predicting
     Construction Contract Surety Bond Claims Using Contractor Financial
     Data." *Journal of Construction Engineering and Management*, Vol.
     120, 405-420.

Shi, J.T.S.; Cheung, S.O and Arditi, D. (2001). "Construction delay
     Computation Method" *Journal of Construction Engineering and
     Management*, Vol. 127, Jan./Feb., 60-65.

Thompson, R.M, Vorster, M.C. and Groton, J.P. (2000). "Innovation to
     Manage Disputes: DRB and NEC, *Journal of Management in
     Engineering*, Vol. 16, No.5, September/October, 51-59.

Trauner, T.J. (1990). "Construction delays", R.S. Means, Kingston, Mass.

Treacy, Thomas, B. (1995). "Use of Alternative Dispute Resolution in the
     Construction Industry", *Journal of Management in Engineering*, Vol.
     11, No. 1, January/February, 58-63.

Tyrril, J. (1992). "Construction Industry Dispute Resolution-a brief
     overview." *Australian Dispute Resolution Journal*, 393), p. 167-183.

Vorster, Mike, C. (1993). "Dispute Review Boards: A Mechanism for Preventing and Resolving Construction Disputes" A Report to the Construction Industry Institute, Virginia Polytechnic Institute and State University.

Wilson, Roy L. (1992). "The Ultimate ADR, AACE International's Professional Practice Guide to Contracts and Claims", *AACE Transactions*.

Yanoviak, John J. (1987). "Improving Project Management for Claims Reduction." AACE International's Professional Practice Guide to Contracts and Claims, *AACE Transactions*.

Yates, J.K. (1993). "Construction decision support system for delay analysis." *J. Construction Engineering and Management.* ASCE, 119(2), 226-244.

Zack, J.G. Jr. (1993). "Claimsmanship: Current perspective." *J. Construction Engineering and Management*, ASCE, 119(3), 480-497.

Zack, J. G. Jr. (1996). "Claims Prevention: Offense Versus Defense." AACE International's Professional Practice Guide to Contracts and Claims, *AACE Transactions*.

Zack J.G. (1997). "Claim prevention: Offence verses defense." *Cost Engineering*, 39(7), 23-28.

Chapter 12 Smart Contracting: a Framework for Better Performance of Contracts and the People Involved

Construction Industry Institute. (1990). "Potential for Construction Industry Improvement" Volume I and II, Document Nos. SD61 and SD62.

Hartman, F.T. (2000). "Smart Trust – a foundation for more effective project management application." *Proceedings of the 15th IPMA World Congress on Project Management (Congress 2000)*, May, London, United Kingdom.

Index